PROBABILITY DEMYSTIFIED

Demystified Series

Advanced Statistics Demystified
Algebra Demystified
Anatomy Demystified
Astronomy Demystified
Biology Demystified
Business Statistics Demystified
Calculus Demystified
Chemistry Demystified
College Algebra Demystified
Differential Equations Demystified
Digital Electronics Demystified
Earth Science Demystified
Electricity Demystified
Electronics Demystified
Everyday Math Demystified
Geometry Demystified
Math Word Problems Demystified
Microbiology Demystified
Physics Demystified
Physiology Demystified
Pre-Algebra Demystified
Precalculus Demystified
Probability Demystified
Project Management Demystified
Robotics Demystified
Statistics Demystified
Trigonometry Demystified

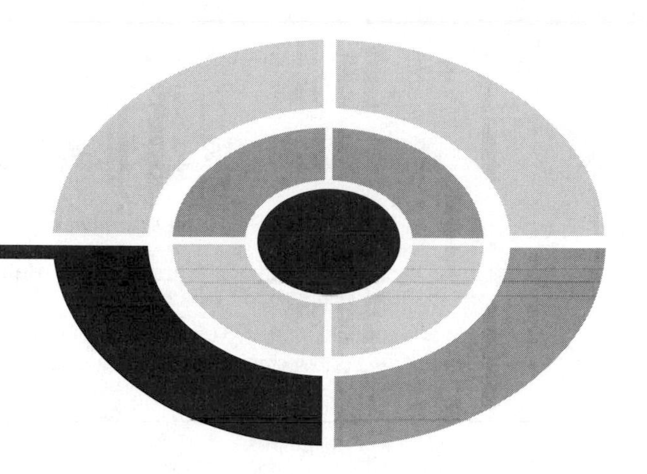

PROBABILITY DEMYSTIFIED

ALLAN G. BLUMAN

McGRAW-HILL
New York Chicago San Francisco Lisbon London
Madrid Mexico City Milan New Delhi San Juan
Seoul Singapore Sydney Toronto

The McGraw-Hill Companies

Library of Congress Cataloging-in-Publication Data

Bluman, Allan G.

Probability demystified / Allan G. Bluman.

p. cm.

Includes index.

ISBN 0-07-144549-8 (acid-free paper)

1. Probabilities. I. Title

QA273.B5929 2005

519.2—dc22 2004065574

4 5 6 7 8 9 0 DOC/DOC 0 1 0 9 8 7

ISBN 0-07-144549-8

The sponsoring editor for this book was Judy Bass and the production supervisor was Pamela A. Pelton. It was set in Times Roman by Keyword Publishing Services Ltd. The art director for the cover was Margaret Webster-Shapiro; the cover designer was Handel Low.

Printed and bound by RR Donnelley.

To all of my teachers, whose examples instilled in me my love of mathematics and teaching.

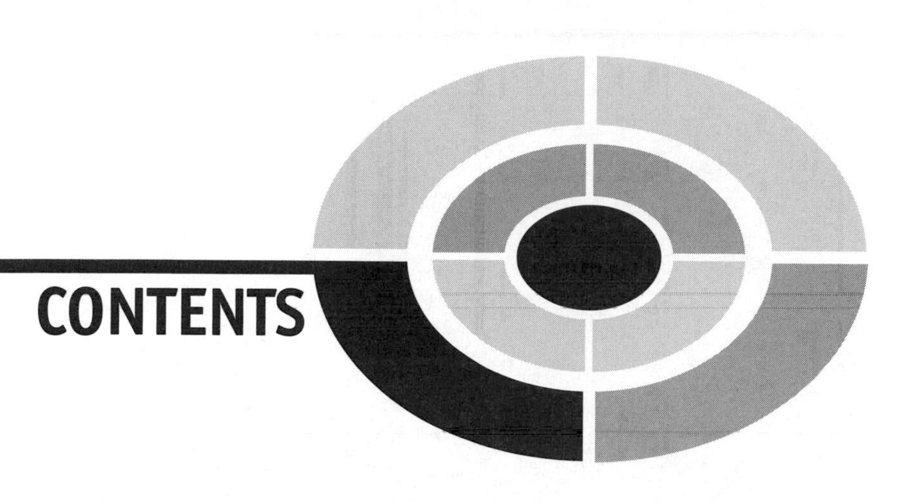

CONTENTS

PREFACE

"*The probable is what usually happens.*" — Aristotle

Probability can be called the mathematics of chance. The theory of probability is unusual in the sense that we cannot predict with certainty the individual outcome of a chance process such as flipping a coin or rolling a die (singular for dice), but we can assign a number that corresponds to the probability of getting a particular outcome. For example, the probability of getting a head when a coin is tossed is 1/2 and the probability of getting a two when a single fair die is rolled is 1/6.

We can also predict with a certain amount of accuracy that when a coin is tossed a large number of times, the ratio of the number of heads to the total number of times the coin is tossed will be close to 1/2.

Probability theory is, of course, used in gambling. Actually, mathematicians began studying probability as a means to answer questions about gambling games. Besides gambling, probability theory is used in many other areas such as insurance, investing, weather forecasting, genetics, and medicine, and in everyday life.

What is this book about?

First let me tell you what this book is **not** about:

- This book is **not** a rigorous theoretical deductive mathematical approach to the concepts of probability.
- This book is **not** a book on how to gamble.

And most important

- This book is **not** a book on how to win at gambling!

This book presents the basic concepts of probability in a simple, straightforward, easy-to-understand way. It does require, however, a knowledge of arithmetic (fractions, decimals, and percents) and a knowledge of basic algebra (formulas, exponents, order of operations, etc.). If you need a review of these concepts, you can consult another of my books in this series entitled *Pre-Algebra Demystified.*

This book can be used to gain a knowledge of the basic concepts of probability theory, either as a self-study guide or as a supplementary textbook for those who are taking a course in probability or a course in statistics that has a section on probability.

The basic concepts of probability are explained in the first two chapters. Then the addition and multiplication rules are explained. Following that, the concepts of odds and expectation are explained. The counting rules are explained in Chapter 6, and they are needed for the binomial and other probability distributions found in Chapters 7 and 8. The relationship between probability and the normal distribution is presented in Chapter 9. Finally, a recent development, the Monte Carlo method of simulation, is explained in Chapter 10. Chapter 11 explains how probability can be used in game theory and Chapter 12 explains how probability is used in actuarial science. Special material on Bayes' Theorem is presented in the Appendix because this concept is somewhat more difficult than the other concepts presented in this book.

In addition to addressing the concepts of probability, each chapter ends with what is called a "Probability Sidelight." These sections cover some of the historical aspects of the development of probability theory or some commentary on how probability theory is used in gambling and everyday life.

I have spent my entire career teaching mathematics at a level that most students can understand and appreciate. I have written this book with the same objective in mind. Mathematical precision, in some cases, has been sacrificed in the interest of presenting probability theory in a simplified way.

Good luck!

Allan G. Bluman

ACKNOWLEDGMENTS

I would like to thank my wife, Betty Claire, for helping me with the preparation of this book and my editor, Judy Bass, for her assistance in its publication. I would also like to thank Carrie Green for her error checking and helpful suggestions.

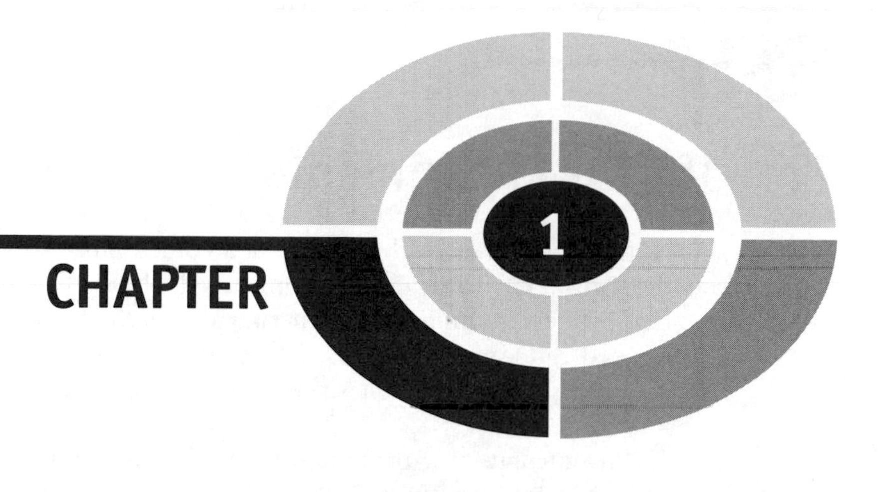

Basic Concepts

Introduction

Probability can be defined as the mathematics of chance. Most people are familiar with some aspects of probability by observing or playing gambling games such as lotteries, slot machines, black jack, or roulette. However, probability theory is used in many other areas such as business, insurance, weather forecasting, and in everyday life.

In this chapter, you will learn about the basic concepts of probability using various devices such as coins, cards, and dice. These devices are not used as examples in order to make you an astute gambler, but they are used because they will help you understand the concepts of probability.

Probability Experiments

Chance processes, such as flipping a coin, rolling a die (singular for dice), or drawing a card at random from a well-shuffled deck are called *probability experiments*. A **probability experiment** is a chance process that leads to well-defined outcomes or results. For example, tossing a coin can be considered a probability experiment since there are two well-defined outcomes—heads and tails.

An **outcome** of a probability experiment is the result of a single trial of a probability experiment. A **trial** means flipping a coin once, or drawing a single card from a deck. A trial could also mean rolling two dice at once, tossing three coins at once, or drawing five cards from a deck at once. A single trial of a probability experiment means to perform the experiment one time.

The set of all outcomes of a probability experiment is called a **sample space.** Some sample spaces for various probability experiments are shown here.

Experiment	Sample Space
Toss one coin	H, T*
Roll a die	1, 2, 3, 4, 5, 6
Toss two coins	HH, HT, TH, TT

*H = heads; T = tails.

Notice that when two coins are tossed, there are four outcomes, not three. Consider tossing a nickel and a dime at the same time. Both coins could fall heads up. Both coins could fall tails up. The nickel could fall heads up and the dime could fall tails up, or the nickel could fall tails up and the dime could fall heads up. The situation is the same even if the coins are indistinguishable.

It should be mentioned that each outcome of a probability experiment occurs at **random**. This means you cannot predict with certainty which outcome will occur when the experiment is conducted. Also, each outcome of the experiment is **equally likely** unless otherwise stated. That means that each outcome has the same probability of occurring.

When finding probabilities, it is often necessary to consider several outcomes of the experiment. For example, when a single die is rolled, you may want to consider obtaining an even number; that is, a two, four, or six. This is called an event. An **event** then usually consists of one or more outcomes of the sample space. (Note: It is sometimes necessary to consider an event which has no outcomes. This will be explained later.)

An event with one outcome is called a **simple event**. For example, a die is rolled and the event of getting a four is a simple event since there is only one way to get a four. When an event consists of two or more outcomes, it is called a **compound event**. For example, if a die is rolled and the event is getting an odd number, the event is a compound event since there are three ways to get an odd number, namely, 1, 3, or 5.

Simple and compound events should not be confused with the number of times the experiment is repeated. For example, if two coins are tossed, the event of getting two heads is a simple event since there is only one way to get two heads, whereas the event of getting a head and a tail in either order is a compound event since it consists of two outcomes, namely head, tail and tail, head.

EXAMPLE: A single die is rolled. List the outcomes in each event:

a. Getting an odd number

b. Getting a number greater than four

c. Getting less than one

SOLUTION:

a. The event contains the outcomes 1, 3, and 5.

b. The event contains the outcomes 5 and 6.

c. When you roll a die, you cannot get a number less than one; hence, the event contains no outcomes.

Classical Probability

Sample spaces are used in **classical probability** to determine the numerical probability that an event will occur. The formula for determining the probability of an event E is

$$P(E) = \frac{\text{number of outcomes contained in the event } E}{\text{total number of outcomes in the sample space}}$$

EXAMPLE: Two coins are tossed; find the probability that both coins land heads up.

SOLUTION:

The sample space for tossing two coins is HH, HT, TH, and TT. Since there are 4 events in the sample space, and only one way to get two heads (HH), the answer is

$$P(\text{HH}) = \frac{1}{4}$$

EXAMPLE: A die is tossed; find the probability of each event:

a. Getting a two
b. Getting an even number
c. Getting a number less than 5

SOLUTION:

The sample space is 1, 2, 3, 4, 5, 6, so there are six outcomes in the sample space.

a. $P(2) = \frac{1}{6}$, since there is only one way to obtain a 2.
b. $P(\text{even number}) = \frac{3}{6} = \frac{1}{2}$, since there are three ways to get an odd number, 1, 3, or 5.
c. $P(\text{number less than } 5) = \frac{4}{6} = \frac{2}{3}$, since there are four numbers in the sample space less than 5.

EXAMPLE: A dish contains 8 red jellybeans, 5 yellow jellybeans, 3 black jellybeans, and 4 pink jellybeans. If a jellybean is selected at random, find the probability that it is

a. A red jellybean
b. A black or pink jellybean
c. Not yellow
d. An orange jellybean

SOLUTION:

There are 8 + 5 + 3 + 4 = 20 outcomes in the sample space.

a. $P(\text{red}) = \frac{8}{20} = \frac{2}{5}$

b. $P(\text{black or pink}) = \frac{3+4}{20} = \frac{7}{20}$

c. $P(\text{not yellow}) = P(\text{red or black or pink}) = \frac{8+3+4}{20} = \frac{15}{20} = \frac{3}{4}$

d. $P(\text{orange}) = \frac{0}{20} = 0$, since there are no orange jellybeans.

Probabilities can be expressed as reduced fractions, decimals, or percents. For example, if a coin is tossed, the probability of getting heads up is $\frac{1}{2}$ or 0.5 or 50%. (Note: Some mathematicians feel that probabilities should be expressed only as fractions or decimals. However, probabilities are often given as percents in everyday life. For example, one often hears, "There is a 50% chance that it will rain tomorrow.")

Probability problems use a certain language. For example, suppose a die is tossed. An event that is specified as "getting at least a 3" means getting a 3, 4, 5, or 6. An event that is specified as "getting at most a 3" means getting a 1, 2, or 3.

Probability Rules

There are certain rules that apply to classical probability theory. They are presented next.

Rule 1: *The probability of any event will always be a number from zero to one.*

This can be denoted mathematically as $0 \leq P(E) \leq 1$. What this means is that all answers to probability problems will be numbers ranging from zero to one. Probabilities cannot be negative nor can they be greater than one.

Also, when the probability of an event is close to zero, the occurrence of the event is relatively unlikely. For example, if the chances that you will win a certain lottery are 0.001 or one in one thousand, you probably won't win, unless of course, you are very "lucky." When the probability of an event is 0.5 or $\frac{1}{2}$, there is a 50–50 chance that the event will happen—the same probability of the two outcomes when flipping a coin. When the probability of an event is close to one, the event is almost sure to occur. For example, if the chance of it snowing tomorrow is 90%, more than likely, you'll see some snow. See Figure 1-1.

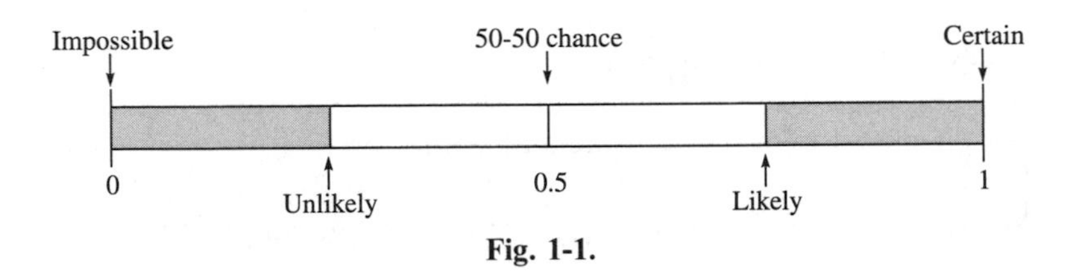

Fig. 1-1.

Rule 2: *When an event cannot occur, the probability will be zero.*

EXAMPLE: A die is rolled; find the probability of getting a 7.

SOLUTION:

Since the sample space is 1, 2, 3, 4, 5, and 6, and there is no way to get a 7, $P(7) = 0$. The event in this case has no outcomes when the sample space is considered.

Rule 3: *When an event is certain to occur, the probability is 1.*

EXAMPLE: A die is rolled; find the probability of getting a number less than 7.

SOLUTION:

Since all outcomes in the sample space are less than 7, the probability is $\frac{6}{6} = 1$.

Rule 4: *The sum of the probabilities of all of the outcomes in the sample space is 1.*

Referring to the sample space for tossing two coins (HH, HT, TH, TT), each outcome has a probability of $\frac{1}{4}$ and the sum of the probabilities of all of the outcomes is

$$\frac{1}{4} + \frac{1}{4} + \frac{1}{4} + \frac{1}{4} = \frac{4}{4} = 1.$$

Rule 5: *The probability that an event will not occur is equal to 1 minus the probability that the event will occur.*

For example, when a die is rolled, the sample space is 1, 2, 3, 4, 5, 6. Now consider the event E of getting a number less than 3. This event consists of the outcomes 1 and 2. The probability of event E is $P(E) = \frac{2}{6} = \frac{1}{3}$. The outcomes in which E will not occur are 3, 4, 5, and 6, so the probability that event E will not occur is $\frac{4}{6} = \frac{2}{3}$. The answer can also be found by substracting from 1, the probability that event E will occur. That is, $1 - \frac{1}{3} = \frac{2}{3}$.

If an event E consists of certain outcomes, then event $\overline{E}$ (E bar) is called the **complement** of event E and consists of the outcomes in the sample space which are not outcomes of event E. In the previous situation, the outcomes in E are 1 and 2. Therefore, the outcomes in $\overline{E}$ are 3, 4, 5, and 6. Now rule five can be stated mathematically as

$$P(\overline{E}) = 1 - P(E).$$

EXAMPLE: If the chance of rain is 0.60 (60%), find the probability that it won't rain.

SOLUTION:

Since $P(E) = 0.60$ and $P(\overline{E}) = 1 - P(E)$, the probability that it won't rain is $1 - 0.60 = 0.40$ or 40%. Hence the probability that it won't rain is 40%.

PRACTICE

1. A box contains a \$1 bill, a \$2 bill, a \$5 bill, a \$10 bill, and a \$20 bill. A person selects a bill at random. Find each probability:

   a. The bill selected is a \$10 bill.
   b. The denomination of the bill selected is more than \$2.
   c. The bill selected is a \$50 bill.
   d. The bill selected is of an odd denomination.
   e. The denomination of the bill is divisible by 5.

2. A single die is rolled. Find each probability:

   a. The number shown on the face is a 2.
   b. The number shown on the face is greater than 2.
   c. The number shown on the face is less than 1.
   d. The number shown on the face is odd.

3. A spinner for a child's game has the numbers 1 through 9 evenly spaced. If a child spins, find each probability:

   a. The number is divisible by 3.
   b. The number is greater than 7.
   c. The number is an even number.

4. Two coins are tossed. Find each probability:

   a. Getting two tails.
   b. Getting at least one head.
   c. Getting two heads.

5. The cards A♥, 2♦, 3♣, 4♥, 5♠, and 6♣ are shuffled and dealt face down on a table. (Hearts and diamonds are red, and clubs and spades are black.) If a person selects one card at random, find the probability that the card is

   a. The 4♥.
   b. A red card.
   c. A club.

6. A ball is selected at random from a bag containing a red ball, a blue ball, a green ball, and a white ball. Find the probability that the ball is

   a. A blue ball.
   b. A red or a blue ball.
   c. A pink ball.

7. A letter is randomly selected from the word "computer." Find the probability that the letter is

   a. A "t".
   b. An "o" or an "m".
   c. An "x".
   d. A vowel.

8. On a roulette wheel there are 38 sectors. Of these sectors, 18 are red, 18 are black, and 2 are green. When the wheel is spun, find the probability that the ball will land on

   a. Red.
   b. Green.

9. A person has a penny, a nickel, a dime, a quarter, and a half-dollar in his pocket. If a coin is selected at random, find the probability that the coin is

   a. A quarter.
   b. A coin whose amount is greater than five cents.
   c. A coin whose denomination ends in a zero.

10. Six women and three men are employed in a real estate office. If a person is selected at random to get lunch for the group, find the probability that the person is a man.

ANSWERS

1. The sample space is \$1, \$2, \$5, \$10, \$20.

   a. $P(\$10) = \frac{1}{5}$.

   b. $P(\text{bill greater than } \$2) = \frac{3}{5}$, since \$5, \$10, and \$20 are greater than \$2.

   c. $P(\$50) = \frac{0}{5} = 0$, since there is no \$50 bill.

   d. $P(\text{bill is odd}) = \frac{2}{5}$, since \$1 and \$5 are odd denominational bills.

   e. $P(\text{number is divisible by } 5) = \frac{3}{5}$, since \$5, \$10, and \$20 are divisible by 5.

2. The sample space is 1, 2, 3, 4, 5, 6.

   a. $P(2) = \frac{1}{6}$, since there is only one 2 in the sample space.

   b. $P(\text{number greater than } 2) = \frac{4}{6} = \frac{2}{3}$, since there are 4 numbers in the sample space greater than 2.

c. P(number less than 1) $= \frac{0}{6} = 0$, since there are no numbers in the sample space less than 1.
d. P(number is an odd number) $= \frac{3}{6} = \frac{1}{2}$, since 1, 3, and 5 are odd numbers.

3. The sample space is 1, 2, 3, 4, 5, 6, 7, 8, 9.

a. P(number divisible by 3) $= \frac{3}{9} = \frac{1}{3}$, since 3, 6, and 9 are divisible by 3.
b. P(number greater than 7) $= \frac{2}{9}$, since 8 and 9 are greater than 7.
c. P(even number) $= \frac{4}{9}$, since 2, 4, 6, and 8 are even numbers.

4. The sample space is HH, HT, TH, TT.

a. P(TT) $= \frac{1}{4}$, since there is only one way to get two tails.
b. P(at least one head) $= \frac{3}{4}$, since there are three ways (HT, TH, HH) to get at least one head.
c. P(HH) $= \frac{1}{4}$, since there is only one way to get two heads.

5. The sample space is A♥, 2♦, 3♣, 4♥, 5♠, 6♣.

a. P(4♥) $= \frac{1}{6}$.
b. P(red card) $= \frac{3}{6} = \frac{1}{2}$, since there are three red cards.
c. P(club) $= \frac{2}{6} = \frac{1}{3}$, since there are two clubs.

6. The sample space is red, blue, green, and white.

a. P(blue) $= \frac{1}{4}$, since there is only one blue ball.
b. P(red or blue) $= \frac{2}{4} = \frac{1}{2}$, since there are two outcomes in the event.
c. P(pink) $= \frac{0}{6} = 0$, since there is no pink ball.

7. The sample space consists of the letters in "computer."

a. P(t) $= \frac{1}{8}$.
b. P(o or m) $= \frac{2}{8} = \frac{1}{4}$.
c. P(x) $= \frac{0}{8} = 0$, since there are no "x"s in the word.
d. P(vowel) $= \frac{3}{8}$, since o, u, and e are the vowels in the word.

8. There are 38 outcomes:

   a. $P(\text{red}) = \frac{18}{38} = \frac{9}{19}.$

   b. $P(\text{green}) = \frac{2}{38} = \frac{1}{19}.$

9. The sample space is 1¢, 5¢, 10¢, 25¢, 50¢.

   a. $P(25¢) = \frac{1}{5}.$

   b. $P(\text{greater than } 5¢) = \frac{3}{5}.$

   c. $P(\text{denomination ends in zero}) = \frac{2}{5}.$

10. The sample space consists of six women and three men.

   $P(\text{man}) = \frac{3}{9} = \frac{1}{3}.$

Empirical Probability

Probabilities can be computed for situations that do not use sample spaces. In such cases, **frequency distributions** are used and the probability is called **empirical probability**. For example, suppose a class of students consists of 4 freshmen, 8 sophomores, 6 juniors, and 7 seniors. The information can be summarized in a frequency distribution as follows:

Rank	Frequency
Freshmen	4
Sophomores	8
Juniors	6
Seniors	7
TOTAL	25

From a frequency distribution, probabilities can be computed using the following formula.

$$P(E) = \frac{\text{frequency of } E}{\text{sum of the frequencies}}$$

Empirical probability is sometimes called **relative frequency** probability.

EXAMPLE: Using the frequency distribution shown previously, find the probability of selecting a junior student at random.

SOLUTION:

Since there are 6 juniors and a total of 25 students, $P(\text{junior}) = \frac{6}{25}$.

Another aspect of empirical probability is that if a large number of subjects (called a **sample**) is selected from a particular group (called a **population**), and the probability of a specific attribute is computed, then when another subject is selected, we can say that the probability that this subject has the same attribute is the same as the original probability computed for the group. For example, a Gallup Poll of 1004 adults surveyed found that 17% of the subjects stated that they considered Abraham Lincoln to be the greatest President of the United States. Now if a subject is selected, the probability that he or she will say that Abraham Lincoln was the greatest president is also 17%.

Several things should be explained here. First of all, the 1004 people constituted a sample selected from a larger group called the population. Second, the exact probability for the population can never be known unless every single member of the group is surveyed. This does not happen in these kinds of surveys since the population is usually very large. Hence, the 17% is only an estimate of the probability. However, if the sample is **representative** of the population, the estimate will usually be fairly close to the exact probability. Statisticians have a way of computing the accuracy (called the margin of error) for these situations. For the present, we shall just concentrate on the probability.

Also, by a representative sample, we mean the subjects of the sample have similar characteristics as those in the population. There are statistical methods to help the statisticians obtain a representative sample. These methods are called sampling methods and can be found in many statistics books.

EXAMPLE: The same study found 7% considered George Washington to be the greatest President. If a person is selected at random, find the probability that he or she considers George Washington to be the greatest President.

SOLUTION:

The probability is 7%.

EXAMPLE: In a sample of 642 people over 25 years of age, 160 had a bachelor's degree. If a person over 25 years of age is selected, find the probability that the person has a bachelor's degree.

SOLUTION:

In this case,

$$P(\text{bachelor's degree}) = \frac{160}{642} = 0.249 \text{ or about } 25\%.$$

EXAMPLE: In the sample study of 642 people, it was found that 514 people have a high school diploma. If a person is selected at random, find the probability that the person does not have a high school diploma.

SOLUTION:

The probability that a person has a high school diploma is

$$P(\text{high school diploma}) = \frac{514}{642} = 0.80 \text{ or } 80\%.$$

Hence, the probability that a person does not have a high school diploma is

$$\begin{aligned} P\ (\text{no high school diploma}) &= 1 - P(\text{high school diploma}) \\ &= 1 - 0.80 = 0.20 \text{ or } 20\%. \end{aligned}$$

Alternate Solution:

If 514 people have a high school diploma, then $642 - 514 = 128$ do not have a high school diploma. Hence

$$P(\text{no high school diploma}) = \frac{128}{642} = 0.199 \text{ or } 20\% \text{ rounded.}$$

Consider another aspect of probability. Suppose a baseball player has a batting average of 0.250. What is the probability that he will get a hit the next time he gets to bat? Although we cannot be sure of the exact probability, we can use 0.250 as an estimate. Since $0.250 = \frac{1}{4}$, we can say that there is a one in four chance that he will get a hit the next time he bats.

PRACTICE

1. A recent survey found that the ages of workers in a factory is distributed as follows:

Age	Number
20–29	18
30–39	27
40–49	36
50–59	16
60 or older	3
Total	100

   If a person is selected at random, find the probability that the person is

   a. 40 or older.
   b. Under 40 years old.
   c. Between 30 and 39 years old.
   d. Under 60 but over 39 years old.

2. In a sample of 50 people, 19 had type O blood, 22 had type A blood, 7 had type B blood, and 2 had type AB blood. If a person is selected at random, find the probability that the person

   a. Has type A blood.
   b. Has type B or type AB blood.
   c. Does not have type O blood.
   d. Has neither type A nor type O blood.

3. In a recent survey of 356 children aged 19–24 months, it was found that 89 ate French fries. If a child is selected at random, find the probability that he or she eats French fries.

4. In a classroom of 36 students, 8 were liberal arts majors and 7 were history majors. If a student is selected at random, find the probability that the student is neither a liberal arts nor a history major.

5. A recent survey found that 74% of those questioned get some of the news from the Internet. If a person is selected at random, find the probability that the person does not get any news from the Internet.

ANSWERS

1. a. $P(\text{40 or older}) = \dfrac{36+16+3}{100} = \dfrac{55}{100} = \dfrac{11}{20}$

   b. $P(\text{under 40}) = \dfrac{18+27}{100} = \dfrac{45}{100} = \dfrac{9}{20}$

   c. $P(\text{between 30 and 39}) = \dfrac{27}{100}$

   d. $P(\text{under 60 but over 39}) = \dfrac{36+16}{100} = \dfrac{52}{100} = \dfrac{13}{25}$

2. The total number of outcomes in this sample space is 50.

   a. $P(\text{A}) = \dfrac{22}{50} = \dfrac{11}{25}$

   b. $P(\text{B or AB}) = \dfrac{7+2}{50} = \dfrac{9}{50}$

   c. $P(\text{not O}) = 1 - \dfrac{19}{50} = \dfrac{31}{50}$

   d. $P(\text{neither A nor O}) = P(\text{AB or B}) = \dfrac{2+7}{50} = \dfrac{9}{50}$

3. $P(\text{French fries}) = \dfrac{89}{356} = \dfrac{1}{4}$

4. $P(\text{neither liberal arts nor history}) = 1 - \dfrac{8+7}{36} = 1 - \dfrac{15}{36} = \dfrac{21}{36} = \dfrac{7}{12}$

5. $P(\text{does not get any news from the Internet}) = 1 - 0.74 = 0.26$

Law of Large Numbers

We know from classical probability that if a coin is tossed one time, we cannot predict the outcome, but the probability of getting a head is $\frac{1}{2}$ and the probability of getting a tail is $\frac{1}{2}$ if everything is fair. But what happens if we toss the coin 100 times? Will we get 50 heads? Common sense tells us that most of the time, we will not get exactly 50 heads, but we should get close to 50 heads. What will happen if we toss a coin 1000 times? Will we get exactly 500 heads? Probably not. However, as the number of tosses increases, the ratio of the number of heads to the total number of tosses will get closer to $\frac{1}{2}$. This phenomenon is known as the **law of large numbers.** This law holds for any type of gambling game such as rolling dice, playing roulette, etc.

Subjective Probability

A third type of probability is called **subjective probability**. Subjective probability is based upon an educated guess, estimate, opinion, or inexact information. For example, a sports writer may say that there is a 30% probability that the Pittsburgh Steelers will be in the Super Bowl next year. Here the sports writer is basing his opinion on subjective information such as the relative strength of the Steelers, their opponents, their coach, etc. Subjective probabilities are used in everyday life; however, they are beyond the scope of this book.

Summary

Probability is the mathematics of chance. There are three types of probability: classical probability, empirical probability, and subjective probability. Classical probability uses sample spaces. A sample space is the set of outcomes of a probability experiment. The range of probability is from 0 to 1. If an event cannot occur, its probability is 0. If an event is certain to occur, its probability is 1. Classical probability is defined as the number of ways (outcomes) the event can occur divided by the total number of outcomes in the sample space.

Empirical probability uses frequency distributions, and it is defined as the frequency of an event divided by the total number of frequencies.

Subjective probability is made by a person's knowledge of the situation and is basically an educated guess as to the chances of an event occurring.

CHAPTER QUIZ

1. Which is not a type of probability?

   a. Classical
   b. Empirical
   c. Subjective
   d. Finite

2. Rolling a die or tossing a coin is called a

   a. Sample experiment
   b. Probability experiment
   c. Infinite experiment
   d. Repeated experiment

3. When an event cannot occur, its probability is

   a. 1
   b. 0
   c. $\frac{1}{2}$
   d. 0.01

4. The set of all possible outcomes of a probability experiment is called the

   a. Sample space
   b. Outcome space
   c. Event space
   d. Experimental space

5. The range of the values a probability can assume is

   a. From 0 to 1
   b. From -1 to $+1$
   c. From 1 to 100
   d. From 0 to $\frac{1}{2}$

6. How many outcomes are there in the sample space when two coins are tossed?

   a. 1
   b. 2
   c. 3
   d. 4

7. The type of probability that uses sample spaces is called

   a. Classical probability
   b. Empirical probability
   c. Subjective probability
   d. Relative probability

8. When an event is certain to occur, its probability is

   a. 0
   b. 1
   c. $\frac{1}{2}$
   d. -1

9. When two coins are tossed, the sample space is

   a. H, T
   b. HH, HT, TT
   c. HH, HT, TH, TT
   d. H, T and HT

10. When a die is rolled, the probability of getting a number greater than 4 is

    a. $\frac{1}{6}$
    b. $\frac{1}{3}$
    c. $\frac{1}{2}$
    d. 1

11. When two coins are tossed, the probability of getting 2 tails is

    a. $\frac{1}{2}$
    b. $\frac{1}{3}$
    c. $\frac{1}{4}$
    d. $\frac{1}{8}$

12. If a letter is selected at random from the word "Mississippi," find the probability that it is an "s."

 a. $\frac{1}{8}$

 b. $\frac{1}{2}$

 c. $\frac{3}{11}$

 d. $\frac{4}{11}$

13. When a die is rolled, the probability of getting an 8 is

 a. $\frac{1}{6}$

 b. 0

 c. 1

 d. $1\frac{1}{2}$

14. In a survey of 180 people, 74 were over the age of 64. If a person is selected at random, what is the probability that the person is over 64?

 a. $\frac{16}{45}$

 b. $\frac{32}{37}$

 c. $\frac{37}{90}$

 d. $\frac{53}{90}$

15. In a classroom of 24 students, there were 20 freshmen. If a student is selected at random, what is the probability that the student is not a freshman?

 a. $\frac{2}{3}$

b. $\frac{5}{6}$

c. $\frac{1}{3}$

d. $\frac{1}{6}$

(The answers to the quizzes are found on pages 242–245.)

Probability Sidelight

BRIEF HISTORY OF PROBABILITY

The concepts of probability are as old as humans. Paintings in tombs excavated in Egypt showed that people played games based on chance as early as 1800 B.C.E. One game was called "Hounds and Jackals" and is similar to the present-day game of "Snakes and Ladders."

Ancient Greeks and Romans made crude dice from various items such as animal bones, stones, and ivory. When some of these items were tested recently, they were found to be quite accurate. These crude dice were also used in fortune telling and divination.

Emperor Claudius (10 BCE–54 CE) is said to have written a book entitled *How To Win at Dice*. He liked playing dice so much that he had a special dice board in his carriage.

No formal study of probability was done until the 16th century when Girolamo Cardano (1501–1576) wrote a book on probability entitled *The Book on Chance and Games*. Cardano was a philosopher, astrologer, physician, mathematician, and gambler. In his book, he also included techniques on how to cheat and how to catch others who are cheating. He is believed to be the first mathematician to formulate a definition of classical probability.

During the mid-1600s, a professional gambler named Chevalier de Mere made a considerable amount of money on a gambling game. He would bet unsuspecting patrons that in four rolls of a die, he could obtain at least one 6. He was so successful at winning that word got around, and people refused to play. He decided to invent a new game in order to keep winning. He would bet patrons that if he rolled two dice 24 times, he would get at least one double 6. However, to his dismay, he began to lose more often than he would win and lost money.

Unable to figure out why he was losing, he asked a renowned mathematician, Blaise Pascal (1623–1662) to study the game. Pascal was a child prodigy when it came to mathematics. At the age of 14, he participated in weekly meetings of the mathematicians of the French Academy. At the age of 16, he invented a mechanical adding machine.

Because of the dice problem, Pascal became interested in studying probability and began a correspondence with a French government official and fellow mathematician, Pierre de Fermat (1601–1665). Together the two were able to solve de Mere's dilemma and formulate the beginnings of probability theory.

In 1657, a Dutch mathematician named Christian Huygens wrote a treatise on the Pascal–Fermat correspondence and introduced the idea of mathematical expectation. (See Chapter 5.)

Abraham de Moivre (1667–1754) wrote a book on probability entitled *Doctrine of Chances* in 1718. He published a second edition in 1738.

Pierre Simon Laplace (1749–1827) wrote a book and a series of supplements on probability from 1812 to 1825. His purpose was to acquaint readers with the theory of probability and its applications, using everyday language. He also stated that the probability that the sun will rise tomorrow is $\frac{1{,}826{,}214}{1{,}826{,}215}$.

Simeon-Denis Poisson (1781–1840) developed the concept of the Poisson distribution. (See Chapter 8.)

Also during the 1800s a mathematician named Carl Friedrich Gauss (1777–1855) developed the concepts of the normal distribution. Earlier work on the normal distribution was also done by de Moivre and Laplace, unknown to Gauss. (See Chapter 9.)

In 1895, the Fey Manufacturing Company of San Francisco invented the first automatic slot machine. These machines consisted of three wheels that were spun when a handle on the side of the machine was pulled. Each wheel contained 20 symbols; however, the number of each type of symbols was not the same on each wheel. For example, the first wheel may have 6 oranges, while the second wheel has 3 oranges, and the third wheel has only one. When a person gets two oranges, the person may think that he has almost won by getting 2 out of 3 equitable symbols, while the real probability of winning is much smaller.

In the late 1940s, two mathematicians, Jon von Neumann and Stanislaw Ulam used a computer to simulate probability experiments. This method is called the Monte Carlo method. (See Chapter 10.)

Today probability theory is used in insurance, gambling, war gaming, the stock market, weather forecasting, and many other areas.

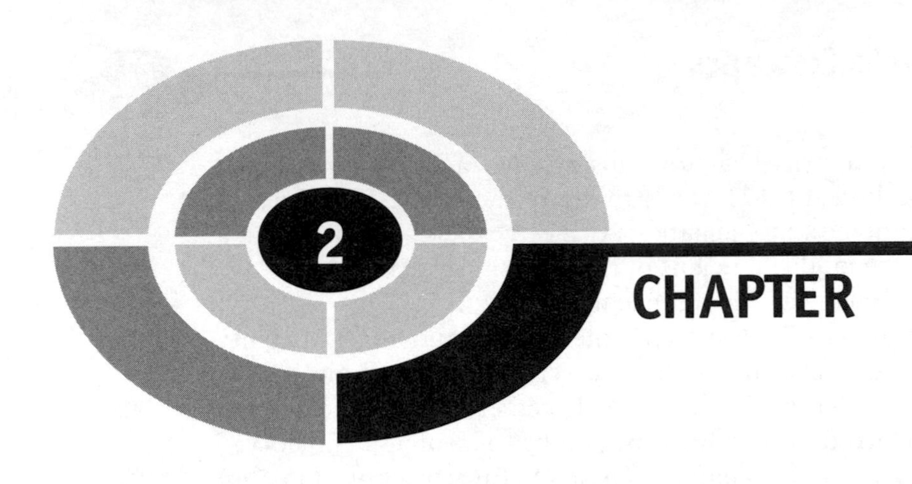

Sample Spaces

Introduction

In order to compute classical probabilities, you need to find the sample space for a probability experiment. In the previous chapter, sample spaces were found by using common sense. In this chapter two specific devices will be used to find sample spaces for probability experiments. They are tree diagrams and tables.

Tree Diagrams

A **tree diagram** consists of branches corresponding to the outcomes of two or more probability experiments that are done in sequence.

In order to construct a tree diagram, use branches corresponding to the outcomes of the first experiment. These branches will emanate from a single point. Then from each branch of the first experiment draw branches that represent the outcomes of the second experiment. You can continue the process for further experiments of the sequence if necessary.

EXAMPLE: A coin is tossed and a die is rolled. Draw a tree diagram and find the sample space.

SOLUTION:

1. Since there are two outcomes (heads and tails for the coin), draw two branches from a single point and label one H for head and the other one T for tail.
2. From each one of these outcomes, draw and label six branches representing the outcomes 1, 2, 3, 4, 5, and 6 for the die.
3. Trace through each branch to find the outcomes of the experiment. See Figure 2-1.

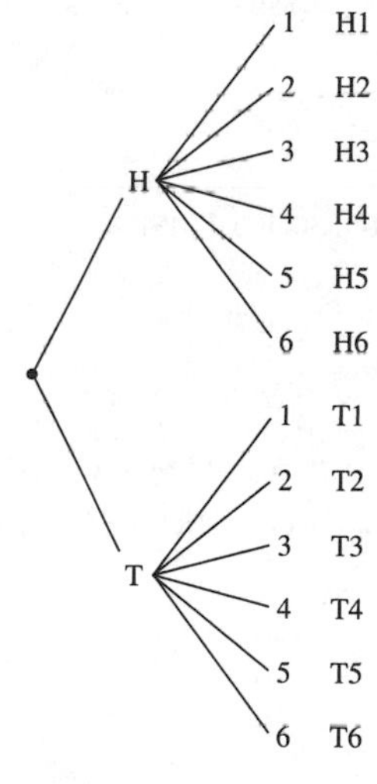

Fig. 2-1.

Hence there are twelve outcomes. They are H1, H2, H3, H4, H5, H6, T1, T2, T3, T4, T5, and T6.

Once the sample space has been found, probabilities for events can be computed.

EXAMPLE: A coin is tossed and a die is rolled. Find the probability of getting

a. A head on the coin and a 3 on the die.
b. A head on the coin.
c. A 4 on the die.

SOLUTION:

a. Since there are 12 outcomes in the sample space and only one way to get a head on the coin and a three on the die,

$$P(\text{H3}) = \frac{1}{12}$$

b. Since there are six ways to get a head on the coin, namely H1, H2, H3, H4, H5, and H6,

$$P(\text{head on the coin}) = \frac{6}{12} = \frac{1}{2}$$

c. Since there are two ways to get a 4 on the die, namely H4 and T4,

$$P(4 \text{ on the die}) = \frac{2}{12} = \frac{1}{6}$$

EXAMPLE: Three coins are tossed. Draw a tree diagram and find the sample space.

SOLUTION:

Each coin can land either heads up (H) or tails up (T); therefore, the tree diagram will consist of three parts and each part will have two branches. See Figure 2-2.

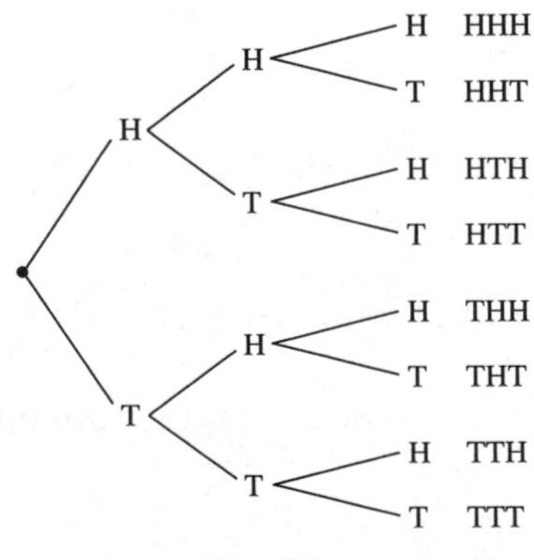

Fig. 2-2.

Hence the sample space is HHH, HHT, HTH, HTT, THH, THT, TTH, TTT.

Once the sample space is found, probabilities can be computed.

EXAMPLE: Three coins are tossed. Find the probability of getting

a. Two heads and a tail in any order.
b. Three heads.
c. No heads.
d. At least two tails.
e. At most two tails.

SOLUTION:

a. There are eight outcomes in the sample space, and there are three ways to get two heads and a tail in any order. They are HHT, HTH, and THH; hence,

$$P(\text{2 heads and a tail}) = \frac{3}{8}$$

b. Three heads can occur in only one way; hence

$$P(\text{HHH}) = \frac{1}{8}$$

c. The event of getting no heads can occur in only one way—namely TTT; hence,

$$P(\text{TTT}) = \frac{1}{8}$$

d. The event of at least two tails means two tails and one head or three tails. There are four outcomes in this event—namely TTH, THT, HTT, and TTT; hence,

$$P(\text{at least two tails}) = \frac{4}{8} = \frac{1}{2}$$

e. The event of getting at most two tails means zero tails, one tail, or two tails. There are seven outcomes in this event—HHH, THH, HTH, HHT, TTH, THT, and HTT; hence,

$$P(\text{at most two tails}) = \frac{7}{8}$$

When selecting more than one object from a group of objects, it is important to know whether or not the object selected is replaced before drawing the second object. Consider the next two examples.

EXAMPLE: A box contains a red ball (R), a blue ball (B), and a yellow ball (Y). Two balls are selected at random in succession. Draw a tree diagram and find the sample space if the first ball is **replaced** before the second ball is selected.

SOLUTION:

There are three ways to select the first ball. They are a red ball, a blue ball, or a yellow ball. Since the first ball is replaced before the second one is selected, there are three ways to select the second ball. They are a red ball, a blue ball, or a yellow ball. The tree diagram is shown in Figure 2-3.

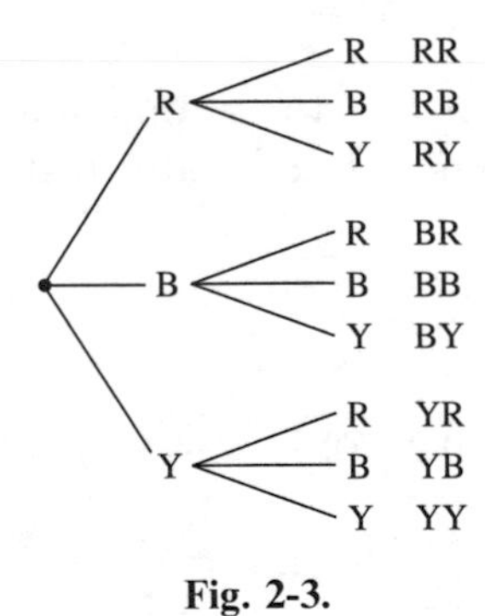

Fig. 2-3.

The sample space consists of nine outcomes. They are RR, RB, RY, BR, BB, BY, YR, YB, YY. Each outcome has a probability of $\frac{1}{9}$.

Now what happens if the first ball is not replaced before the second ball is selected?

EXAMPLE: A box contains a red ball (R), a blue ball (B), and a yellow ball (Y). Two balls are selected at random in succession. Draw a tree diagram and find the sample space if the first ball is **not replaced** before the second ball is selected.

SOLUTION:

There are three outcomes for the first ball. They are a red ball, a blue ball, or a yellow ball. Since the first ball is not replaced before the second ball is drawn, there are only two outcomes for the second ball, and these outcomes depend on the color of the first ball selected. If the first ball selected is blue, then the second ball can be either red or yellow, etc. The tree diagram is shown in Figure 2-4.

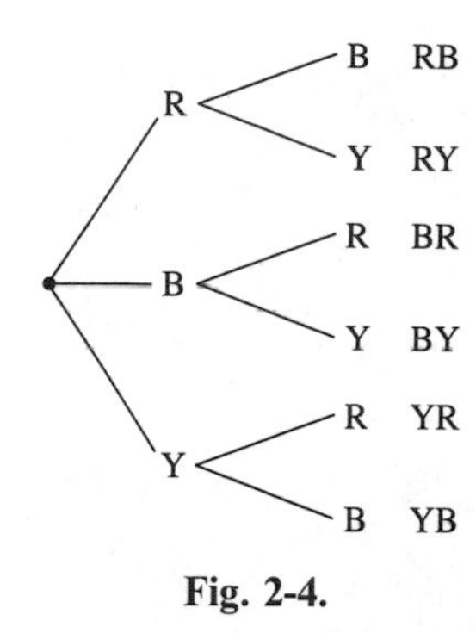

Fig. 2-4.

The sample space consists of six outcomes, which are RB, RY, BR, BY, YR, YB. Each outcome has a probability of $\frac{1}{6}$.

PRACTICE

1. If the possible blood types are A, B, AB, and O, and each type can be Rh^+ or Rh^-, draw a tree diagram and find all possible blood types.
2. Students are classified as male (M) or female (F), freshman (Fr), sophomore (So), junior (Jr), or senior (Sr), and full-time (Ft) or part-time (Pt). Draw a tree diagram and find all possible classifications.
3. A box contains a $1 bill, a $5 bill, and a $10 bill. Two bills are selected in succession **with** replacement. Draw a tree diagram and find the sample space. Find the probability that the total amount of money selected is

   a. $6.
   b. Greater than $10.
   c. Less than $15.

4. Draw a tree diagram and find the sample space for the genders of the children in a family consisting of 3 children. Assume the genders are equiprobable. Find the probability of

   a. Three girls.
   b. Two boys and a girl in any order.
   c. At least two boys.

5. A box contains a white marble (W), a blue marble (B), and a green marble (G). Two marbles are selected **without** replacement. Draw a tree diagram and find the sample space. Find the probability that one marble is white.

ANSWERS

1.

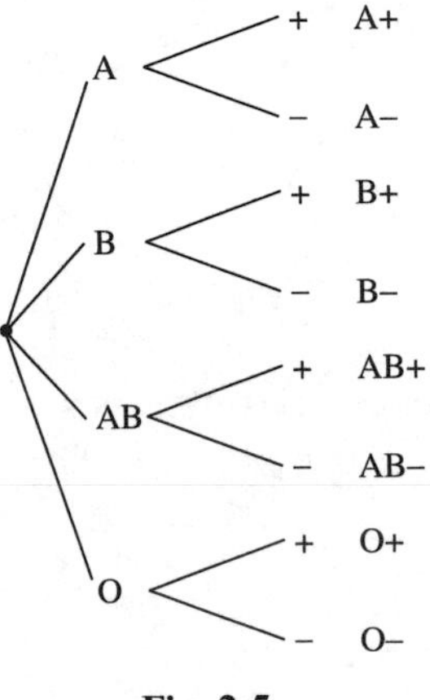

Fig. 2-5.

2.

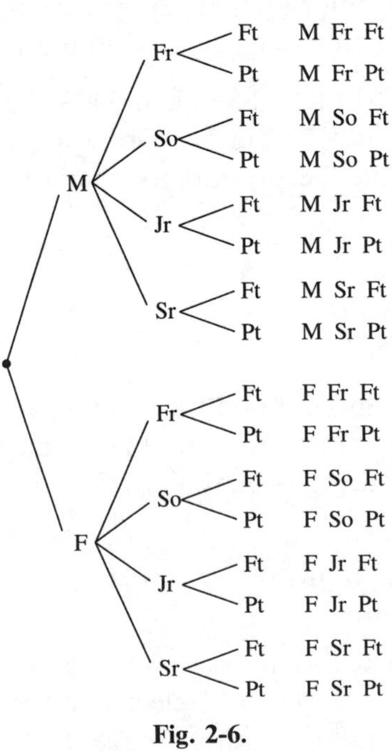

Fig. 2-6.

3.

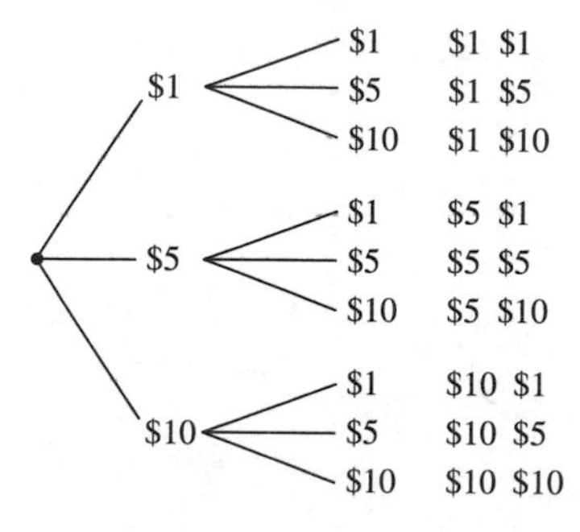

Fig. 2-7.

There are nine outcomes in the sample space.

a. $P(\$6) = \frac{2}{9}$ since $\$1 + \$5$, and $\$5 + \$1$ equal $6.

b. $P(\text{greater than } \$10) = \frac{5}{9}$ since there are five ways to get a sum greater than $10.

c. $P(\text{less than } \$15) = \frac{6}{9} = \frac{2}{3}$ since there are six ways to get a sum lesser than $15.

4.

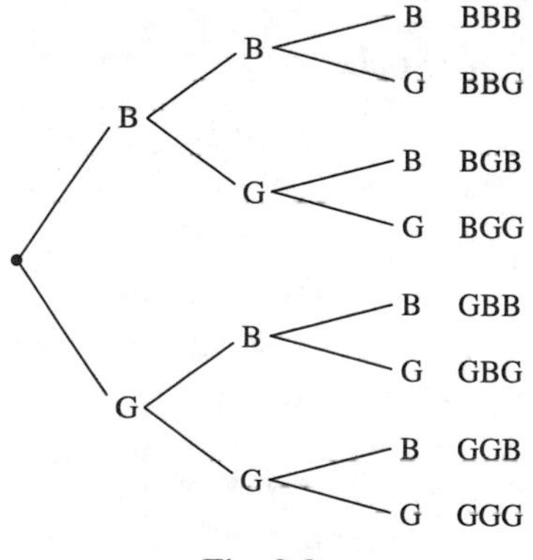

Fig. 2-8.

There are eight outcomes in the sample space.

a. $P(3 \text{ girls}) = \frac{1}{8}$ since three girls is GGG.

b. $P(2 \text{ boys and one girl in any order}) = \frac{3}{8}$ since there are three ways to get two boys and one girl in any order. They are BBG, BGB, and GBB.

c. $P(\text{at least 2 boys}) = \frac{4}{8} = \frac{1}{2}$ since at least two boys means two or three boys. The outcomes are BBG, BGB, GBB, and BBB.

5.

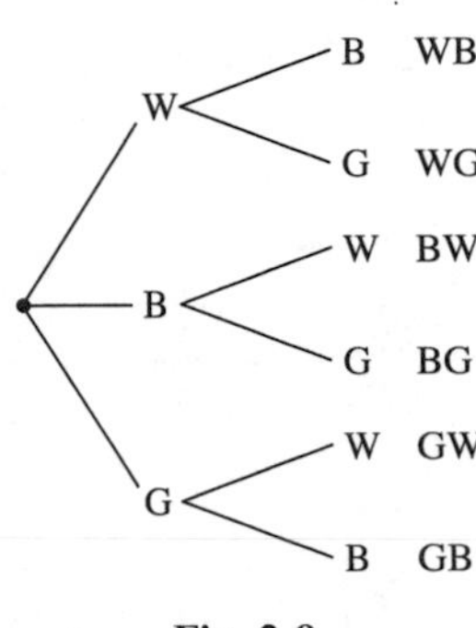

Fig. 2-9.

The probability that one marble is white is $\frac{4}{6} = \frac{2}{3}$ since the outcomes are WB, WG, BW, and GW.

Tables

Another way to find a sample space is to use a table.

EXAMPLE: Find the sample space for selecting a card from a standard deck of 52 cards.

SOLUTION:

There are four suits—hearts and diamonds, which are red, and spades and clubs, which are black. Each suit consists of 13 cards—ace through king. Hence, the sample space can be shown using a table. See Figure 2-10.

A	2	3	4	5	6	7	8	9	10	J	Q	K
♥	♥	♥	♥	♥	♥	♥	♥	♥	♥	♥	♥	♥
A	2	3	4	5	6	7	8	9	10	J	Q	K
♦	♦	♦	♦	♦	♦	♦	♦	♦	♦	♦	♦	♦
A	2	3	4	5	6	7	8	9	10	J	Q	K
♠	♠	♠	♠	♠	♠	♠	♠	♠	♠	♠	♠	♠
A	2	3	4	5	6	7	8	9	10	J	Q	K
♣	♣	♣	♣	♣	♣	♣	♣	♣	♣	♣	♣	♣

Fig. 2-10.

Face cards are kings, queens, and jacks.

Once the sample space is found, probabilities for events can be computed.

EXAMPLE: A single card is drawn at random from a standard deck of cards. Find the probability that it is

a. The 4 of diamonds.
b. A queen.
c. A 5 or a heart.

SOLUTION:

a. The sample space consists of 52 outcomes and only one outcome is the four of diamonds; hence,

$$P(4\blacklozenge) = \frac{1}{52}$$

b. Since there are four queens (one of each suit),

$$P(Q) = \frac{4}{52} = \frac{1}{13}$$

c. In this case, there are 13 hearts and 4 fives; however, the 5♥ has been counted twice, so the number of ways to get a 5 or a heart is $13 + 4 - 1 = 16$. Hence,

$$P(5 \text{ or } \heartsuit) = \frac{16}{52} = \frac{4}{13}.$$

A table can be used for the sample space when two dice are rolled. Since the first die can land in 6 ways and the second die can land in 6 ways, there are 6×6 or 36 outcomes in the sample space. It does not matter whether the two dice are of the same color or different color. The sample space is shown in Figure 2-11.

	Die 2					
Die 1	1	2	3	4	5	6
1	(1, 1)	(2, 1)	(3, 1)	(4, 1)	(5, 1)	(6, 1)
2	(1, 2)	(2, 2)	(3, 2)	(4, 2)	(5, 2)	(6, 2)
3	(1, 3)	(2, 3)	(3, 3)	(4, 3)	(5, 3)	(6, 3)
4	(1, 4)	(2, 4)	(3, 4)	(4, 4)	(5, 4)	(6, 4)
5	(1, 5)	(2, 5)	(3, 5)	(4, 5)	(5, 5)	(6, 5)
6	(1, 6)	(2, 6)	(3, 6)	(4, 6)	(5, 6)	(6, 6)

Fig. 2-11.

Notice that the sample space consists of ordered pairs of numbers. The outcome (4, 2) means that a 4 was obtained on the first die and a 2 was obtained on the second die. The sum of the spots on the faces in this case is $4+2=6$. Probability problems involving rolling two dice can be solved using the sample space shown in Figure 2-11.

EXAMPLE: When two dice are rolled, find the probability of getting a sum of nine.

SOLUTION:

There are four ways of rolling a nine. They are (6, 3), (5, 4), (4, 5), and (3, 6). The sample space consists of 36 outcomes. Hence,

$$P(9) = \frac{4}{36} = \frac{1}{9}$$

EXAMPLE: When two dice are rolled, find the probability of getting doubles.

SOLUTION:

There are six ways to get doubles. They are (1, 1), (2, 2), (3, 3), (4, 4), (5, 5), and (6, 6); hence

$$P(\text{doubles}) = \frac{6}{36} = \frac{1}{6}$$

EXAMPLE: When two dice are rolled, find the probability of getting a sum less than five.

SOLUTION:

A sum less than five means a sum of four, three, or two. There are three ways of getting a sum of four. They are (3, 1), (2, 2), and (1, 3). There are two ways of getting a sum of three. They are (2, 1), and (1, 2). There is one way of getting a sum of two. It is (1, 1). The total number of ways of getting a sum less than five is $3+2+1=6$. Hence,

$$P(\text{sum less than 6}) = \frac{6}{36} = \frac{1}{6}$$

EXAMPLE: When two dice are rolled, find the probability that one of the numbers is a 6.

SOLUTION:

There are 11 outcomes that contain a 6. They are (1, 6), (2, 6), (3, 6), (4, 6), (5, 6), (6, 6), (6, 5), (6, 4), (6, 3), (6, 2), and (6, 1). Hence,

$$P(\text{one of the numbers is a six}) = \frac{11}{36}$$

PRACTICE

1. When a card is selected at random, find the probability of getting
   a. A 9.
   b. The ace of diamonds.
   c. A club.
2. When a card is selected at random from a deck, find the probability of getting
   a. A black card.
   b. A red queen.
   c. A heart or a spade.
3. When a card is selected at random from a deck, find the probability of getting
   a. A diamond or a face card.
   b. A club or an 8.
   c. A red card or a 6.
4. When two dice are rolled, find the probability of getting
   a. A sum of 7.
   b. A sum greater than 8.
   c. A sum less than or equal to 5.
5. When two dice are rolled, find the probability of getting
   a. A 5 on one or both dice.
   b. A sum greater than 12.
   c. A sum less than 13.

ANSWERS

1. There are 52 outcomes in the sample space.
   a. There are four 9s, so

$$P(9) = \frac{4}{52} = \frac{1}{13}$$

   b. There is only one ace of diamonds, so

$$P(\text{A}\blacklozenge) = \frac{1}{52}$$

   c. There are 13 clubs, so

$$P(\clubsuit) = \frac{13}{52} = \frac{1}{4}$$

2. There are 52 outcomes in the sample space.
   a. There are 26 black cards: they are 13 clubs and 13 spades, so

$$P(\text{black card}) = \frac{26}{52} = \frac{1}{2}$$

   b. There are two red queens: they are the queen of diamonds and the queen of hearts, so

$$P(\text{red queen}) = \frac{2}{52} = \frac{1}{26}$$

   c. There are 13 hearts and 13 spades, so

$$P(\heartsuit \text{ or } \clubsuit) = \frac{26}{52} = \frac{1}{2}$$

3. There are 52 outcomes in the sample space.
   a. There are 13 diamonds and 12 face cards, but the jack, queen, and king of diamonds have been counted twice, so

$$P(\text{diamond or face card}) = \frac{13 + 12 - 4}{52} = \frac{21}{52}$$

   b. There are 13 clubs and four 8s, but the 8 of clubs has been counted twice, so

$$P(\clubsuit \text{ or an } 8) = \frac{13 + 4 - 1}{52} = \frac{16}{52} = \frac{4}{13}$$

c. There are 26 red cards and four 6s, but the 6 of hearts and the 6 of diamonds have been counted twice, so

$$P(\text{red card or } 6) = \frac{26 + 4 - 2}{52} = \frac{28}{52} = \frac{7}{13}$$

4. There are 36 outcomes in the sample space.
   a. There are six ways to get a sum of seven. They are (1, 6), (2, 5), (3, 4), (4, 3), (5, 2) and (6, 1); hence,

$$P(\text{sum of } 7) = \frac{6}{36} = \frac{1}{6}$$

   b. A sum greater than 8 means a sum of 9, 10, 11, 12, so

$$P(\text{sum greater than } 8) = \frac{10}{36} = \frac{5}{18}$$

   c. A sum less than or equal to five means a sum of five, four, three or two. There are ten ways to get a sum less than or equal to five; hence,

$$P(\text{sum less than or equal to five}) = \frac{10}{36} = \frac{5}{18}$$

5. There are 36 outcomes in the sample space.
   a. There are 11 ways to get a 5 on one or both dice. They are (1, 5), (2, 5), (3, 5), (4, 5), (5, 5), (6, 5), (5, 6), (5, 4), (5, 3) (5, 2), and (5, 1); hence,

$$P(5 \text{ on one or both dice}) = \frac{11}{36}$$

   b. There are 0 ways to get a sum greater than 12; hence,

$$P(\text{sum greater than } 12) = \frac{0}{36} = 0$$

   The event is impossible.

   c. Since all sums are less than 13 when two dice are rolled, there are 36 ways to get a sum less than 13; hence,

$$P(\text{sum less than } 13) = \frac{36}{36} = 1$$

   The event is certain.

Summary

Two devices can be used to represent sample spaces. They are tree diagrams and tables.

A tree diagram can be used to determine the outcome of a probability experiment. A tree diagram consists of branches corresponding to the outcomes of two or more probability experiments that are done in sequence.

Sample spaces can also be represented by using tables. For example, the outcomes when selecting a card from an ordinary deck can be represented by a table. When two dice are rolled, the 36 outcomes can be represented by using a table. Once a sample space is found, probabilities can be computed for specific events.

CHAPTER QUIZ

1. When a coin is tossed and then a die is rolled, the probability of getting a tail on the coin and an odd number on the die is

   a. $\frac{1}{2}$

   b. $\frac{1}{4}$

   c. $\frac{3}{4}$

   d. $\frac{1}{12}$

2. When a coin is tossed and a die is rolled, the probability of getting a head and a number less than 5 on the die is

   a. $\frac{1}{3}$

   b. $\frac{2}{3}$

   c. $\frac{1}{2}$

   d. $\frac{5}{6}$

3. When three coins are tossed, the probability of getting at least one tail is

   a. $\frac{3}{8}$

   b. $\frac{1}{8}$

   c. $\frac{7}{8}$

   d. $\frac{5}{8}$

4. When three coins are tossed, the probability of getting two or more heads is

   a. $\frac{3}{8}$

   b. $\frac{1}{8}$

   c. $\frac{1}{2}$

   d. $\frac{7}{8}$

5. A box contains a penny, a nickel, a dime, and a quarter. If two coins are selected without replacement, the probability of getting an amount greater than 11¢ is

   a. $\frac{5}{72}$

   b. $\frac{2}{3}$

   c. $\frac{3}{4}$

   d. $\frac{5}{6}$

6. A bag contains a red bead, a green bead, and a blue bead. If a bead is selected and its color noted, and then it is replaced and another bead is selected, the probability that both beads will be of the same color is

   a. $\frac{1}{8}$

   b. $\frac{3}{4}$

   c. $\frac{1}{16}$

   d. $\frac{1}{3}$

7. A card is selected at random from an ordinary deck of 52 cards. The probability that the 7 of diamonds is selected is

   a. $\frac{1}{13}$

   b. $\frac{1}{4}$

   c. $\frac{1}{52}$

   d. $\frac{1}{26}$

8. A card is selected at random from a deck of 52 cards. The probability that it is a 7 is

   a. $\frac{1}{4}$

   b. $\frac{1}{52}$

   c. $\frac{7}{52}$

   d. $\frac{1}{13}$

9. A card is drawn from an ordinary deck of 52 cards. The probability that it is a spade is

 a. $\frac{1}{4}$

 b. $\frac{1}{13}$

 c. $\frac{1}{52}$

 d. $\frac{1}{26}$

10. A card is drawn from an ordinary deck of 52 cards. The probability that it is a 9 or a club is

 a. $\frac{17}{52}$

 b. $\frac{5}{8}$

 c. $\frac{4}{13}$

 d. $\frac{3}{4}$

11. A card is drawn from an ordinary deck of 52 cards. The probability that it is a face card is

 a. $\frac{3}{52}$

 b. $\frac{1}{4}$

 c. $\frac{9}{13}$

 d. $\frac{3}{13}$

12. Two dice are rolled. The probability that the sum of the spots on the faces will be nine is

    a. $\frac{1}{9}$

    b. $\frac{5}{36}$

    c. $\frac{1}{6}$

    d. $\frac{3}{13}$

13. Two dice are rolled. The probability that the sum of the spots on the faces is greater than seven is

    a. $\frac{2}{3}$

    b. $\frac{7}{36}$

    c. $\frac{3}{4}$

    d. $\frac{5}{12}$

14. Two dice are rolled. The probability that one or both numbers on the faces will be 4 is

    a. $\frac{1}{3}$

    b. $\frac{4}{13}$

    c. $\frac{11}{36}$

    d. $\frac{1}{6}$

15. Two dice are rolled. The probability that the sum of the spots on the faces will be even is

 a. $\frac{3}{4}$

 b. $\frac{5}{6}$

 c. $\frac{1}{2}$

 d. $\frac{1}{6}$

Probability Sidelight

HISTORY OF DICE AND CARDS

Dice are one of the earliest known gambling devices used by humans. They have been found in ancient Egyptian tombs and in the prehistoric caves of people in Europe and America. The first dice were made from animal bones—namely the astragalus or the heel bone of a hoofed animal. These bones are very smooth and easily carved. The astragalus had only four sides as opposed to modern cubical dice that have six sides. The astragalus was used for fortune telling, gambling, and board games.

By 3000 B.C.E. the Egyptians had devised many board games. Ancient tomb paintings show pharaohs playing board games, and a game similar to today's "Snakes and Ladders" was found in an Egyptian tomb dating to 1800 B.C.E. Eventually crude cubic dice evolved from the astragalus. The dice were first made from bones, then clay, wood, and finally polished stones. Dots were used instead of numbers since writing numbers was very complicated at that time.

It was thought that the outcomes of rolled dice were controlled by the gods that the people worshipped. As one story goes, the Romans incorrectly reasoned that there were three ways to get a sum of seven when two dice are rolled. They are 6 and 1, 5 and 2, and 4 and 3. They also reasoned incorrectly that there were three ways to get a sum of six: 5 and 1, 4 and 2, and 3 and 3.

They knew from gambling with dice that a sum of seven appeared more than a sum of six. They believed that the reason was that the gods favored the number seven over the number six, since seven at the time was considered a "lucky number." Furthermore, they even "loaded" dice so that the faces showing one and six occurred more often than other faces would, if the dice were fair.

It is interesting to note that on today's dice, the numbers on the opposite faces sum to seven. That is, 4 is opposite 3, 2 is opposite 5, and 6 is opposite 1. This was not always true. Early dice showed 1 opposite 2, 3 opposite 4, and 5 opposite 6. The changeover to modern configuration is believed to have occurred in Egypt.

Many of the crude dice have been tested and found to be quite accurate. Actually mathematicians began to study the outcomes of dice only around the 16th century. The great astronomer Galileo Galilei is usually given the credit for figuring out that when three dice are rolled, there are 216 total outcomes, and that a sum of 10 and 11 is more probable than a sum of 9 and 12. This fact was known intuitively by gamblers long before this time.

Today, dice are used in many types of gambling games and many types of board games. Where would we be today without the game of Monopoly?

It is thought that playing cards evolved from long wooden sticks that had various markings and were used by early fortunetellers and gamblers in the Far East. When the Chinese invented paper over 2000 years ago, people marked long thin strips of paper and used them instead of wooden sticks.

Paper "cards" first appeared in Europe around 1300 and were widely used in most of the European countries. Some decks contained 17 cards; others had 22 cards. The early cards were hand-painted and quite expensive to produce. Later stencils were used to cut costs.

The markings on the cards changed quite often. Besides the four suits commonly used today, early decks of cards had 5 or 6 suits and used other symbols such as coins, flowers, and leaves.

The first cards to be manufactured in the United States were made by Jazaniah Ford in the late 1700s. His company lasted over 50 years. The first book on gambling published in the United States was an edition of *Hoyle's Games*, which was printed in 1796.

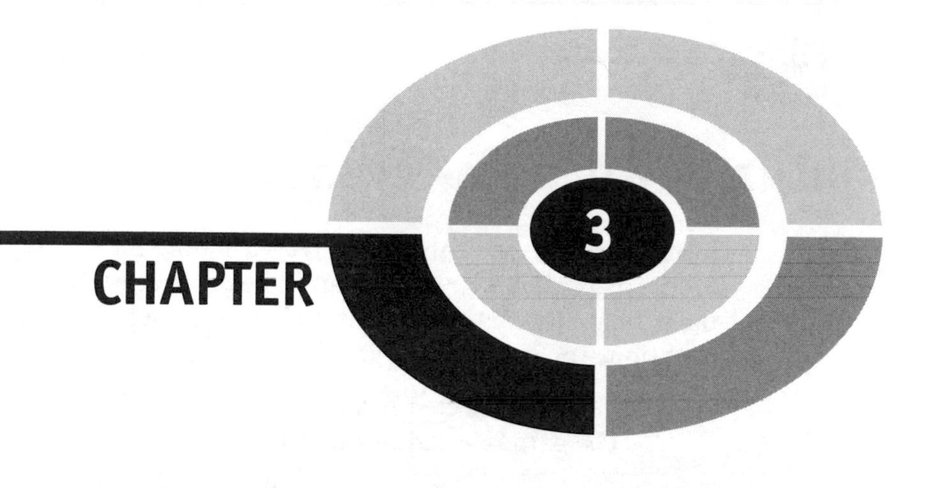

CHAPTER

The Addition Rules

Introduction

In this chapter, the theory of probability is extended by using what are called the **addition rules**. Here one is interested in finding the probability of one event **or** another event occurring. In these situations, one must consider whether or not both events have common outcomes. For example, if you are asked to find the probability that you will get three oranges or three cherries on a slot machine, you know that these two events cannot occur at the same time if the machine has only three windows. In another situation you may be asked to find the probability of getting an odd number or a number less than 500 on a daily three-digit lottery drawing. Here the events have common outcomes. For example, the number 451 is an odd number and a number less than 500. The two addition rules will enable you to solve these kinds of problems as well as many other probability problems.

Mutually Exclusive Events

Many problems in probability involve finding the probability of two or more events. For example, when a card is selected at random from a deck, what is the probability that the card is a king or a queen? In this case, there are two situations to consider. They are:

1. The card selected is a king
2. The card selected is a queen

Now consider another example. When a card is selected from a deck, find the probability that the card is a king or a diamond.

In this case, there are three situations to consider:

1. The card is a king
2. The card is a diamond
3. The card is a king and a diamond. That is, the card is the king of diamonds.

The difference is that in the first example, a card cannot be both a king and a queen at the same time, whereas in the second example, it is possible for the card selected to be a king and a diamond at the same time. In the first example, we say the two events are **mutually exclusive**. In the second example, we say the two events are **not** mutually exclusive. Two events then are **mutually exclusive** if they cannot occur at the same time. In other words, the events have no common outcomes.

EXAMPLE: Which of these events are mutually exclusive?

a. Selecting a card at random from a deck and getting an ace or a club
b. Rolling a die and getting an odd number or a number less than 4
c. Rolling two dice and getting a sum of 7 or 11
d. Selecting a student at random who is full-time or part-time
e. Selecting a student who is a female or a junior

SOLUTION:

a. No. The ace of clubs is an outcome of both events.
b. No. One and three are common outcomes.
c. Yes
d. Yes
e. No. A female student who is a junior is a common outcome.

Addition Rule I

The probability of two or more events occurring can be determined by using the **addition rules.** The first rule is used when the events are mutually exclusive.

Addition Rule I: When two events are mutually exclusive,

$$P(A \text{ or } B) = P(A) + P(B)$$

EXAMPLE: When a die is rolled, find the probability of getting a 2 or a 3.

SOLUTION:

As shown in Chapter 1, the problem can be solved by looking at the sample space, which is 1, 2, 3, 4, 5, 6. Since there are 2 favorable outcomes from 6 outcomes, $P(2 \text{ or } 3) = \frac{2}{6} = \frac{1}{3}$. Since the events are mutually exclusive, addition rule 1 also can be used:

$$P(2 \text{ or } 3) = P(2) + P(3) = \frac{1}{6} + \frac{1}{6} = \frac{2}{6} = \frac{1}{3}$$

EXAMPLE: In a committee meeting, there were 5 freshmen, 6 sophomores, 3 juniors, and 2 seniors. If a student is selected at random to be the chairperson, find the probability that the chairperson is a sophomore or a junior.

SOLUTION:

There are 6 sophomores and 3 juniors and a total of 16 students.

$$P(\text{sophomore or junior}) = P(\text{sophomore}) + P(\text{junior}) = \frac{6}{16} + \frac{3}{16} = \frac{9}{16}$$

EXAMPLE: A card is selected at random from a deck. Find the probability that the card is an ace or a king.

SOLUTION:

$$P(\text{ace or king}) = P(\text{ace}) + P(\text{king}) = \frac{4}{52} + \frac{4}{52} = \frac{8}{52} = \frac{2}{13}$$

The word **or** is the key word, and it means one event occurs or the other event occurs.

PRACTICE

1. In a box there are 3 red pens, 5 blue pens, and 2 black pens. If a person selects a pen at random, find the probability that the pen is

   a. A blue or a red pen.
   b. A red or a black pen.

2. A small automobile dealer has 4 Buicks, 7 Fords, 3 Chryslers, and 6 Chevrolets. If a car is selected at random, find the probability that it is

   a. A Buick or a Chevrolet.
   b. A Chrysler or a Chevrolet.

3. In a model railroader club, 23 members model HO scale, 15 members model N scale, 10 members model G scale, and 5 members model O scale. If a member is selected at random, find the probability that the member models

   a. N or G scale.
   b. HO or O scale.

4. A package of candy contains 8 red pieces, 6 white pieces, 2 blue pieces, and 4 green pieces. If a piece is selected at random, find the probability that it is

   a. White or green.
   b. Blue or red.

5. On a bookshelf in a classroom there are 6 mathematics books, 5 reading books, 4 science books, and 10 history books. If a student selects a book at random, find the probability that the book is

   a. A history book or a mathematics book.
   b. A reading book or a science book.

ANSWERS

1. a. $P(\text{blue or red}) = P(\text{blue}) + P(\text{red}) = \frac{5}{10} + \frac{3}{10} = \frac{8}{10} = \frac{4}{5}$

   b. $P(\text{red or black}) = P(\text{red}) + P(\text{black}) = \frac{3}{10} + \frac{2}{10} = \frac{5}{10} = \frac{1}{2}$

2. a. $P(\text{Buick or Chevrolet}) = P(\text{Buick}) + P(\text{Chevrolet})$

$$= \frac{4}{20} + \frac{6}{20} = \frac{10}{20} = \frac{1}{2}$$

b. $P(\text{Chrysler or Chevrolet}) = P(\text{Chrysler}) + P(\text{Chevrolet})$

$$= \frac{3}{20} + \frac{6}{20} = \frac{9}{20}$$

3. a. $P(\text{N or G}) = P(\text{N}) + P(\text{G}) = \frac{15}{53} + \frac{10}{53} = \frac{25}{53}$

b. $P(\text{HO or O}) = P(\text{HO}) + P(\text{O}) = \frac{23}{53} + \frac{5}{53} = \frac{28}{53}$

4. a. $P(\text{white or green}) = P(\text{white}) + P(\text{green}) = \frac{6}{20} + \frac{4}{20} = \frac{10}{20} = \frac{1}{2}$

b. $P(\text{blue or red}) = P(\text{blue}) + P(\text{red}) = \frac{2}{20} + \frac{8}{20} = \frac{10}{20} = \frac{1}{2}$

5. a. $P(\text{history or math}) = P(\text{history}) + P(\text{math}) = \frac{10}{25} + \frac{6}{25} = \frac{16}{25}$

b. $P(\text{reading or science}) = P(\text{reading}) + P(\text{science}) = \frac{5}{25} + \frac{4}{25} = \frac{9}{25}$

Addition Rule II

When two events are not mutually exclusive, you need to add the probabilities of each of the two events and subtract the probability of the outcomes that are common to both events. In this case, addition rule II can be used.

Addition Rule II: If A and B are two events that are not mutually exclusive, then $P(A \text{ or } B) = P(A) + P(B) - P(A \text{ and } B)$, where A and B means the number of outcomes that event A and event B have in common.

EXAMPLE: A card is selected at random from a deck of 52 cards. Find the probability that it is a 6 or a diamond.

SOLUTION:

Let A = the event of getting a 6; then $P(A) = \frac{4}{52}$ since there are four 6s. Let B = the event of getting a diamond; then $P(B) = \frac{13}{52}$ since there are 13 diamonds. Since there is one card that is both a 6 and a diamond (i.e., the 6 of diamonds), $P(A \text{ and } B) = \frac{1}{52}$. Hence,

$$P(A \text{ or } B) = P(A) + P(B) - P(A \text{ and } B) = \frac{4}{52} + \frac{13}{52} - \frac{1}{52} = \frac{16}{52} = \frac{4}{13}$$

EXAMPLE: A die is rolled. Find the probability of getting an even number or a number less than 4.

SOLUTION:

Let A = an even number; then $P(A) = \frac{3}{6}$ since there are 3 even numbers—2, 4, and 6. Let B = a number less than 4; then $P(B) = \frac{3}{6}$ since there are 3 numbers less than 4—1, 2, and 3. Let (A and B) = even numbers less than 4 and $P(A \text{ and } B) = \frac{1}{6}$ since there is one even number less than 4—namely 2. Hence,

$$P(A \text{ or } B) = P(A) + P(B) - P(A \text{ and } B) = \frac{3}{6} + \frac{3}{6} - \frac{1}{6} = \frac{5}{6}$$

The results of both these examples can be verified by using sample spaces and classical probability.

EXAMPLE: Two dice are rolled; find the probability of getting doubles or a sum of 8.

SOLUTION:

Let A = getting doubles; then $P(A) = \frac{6}{36}$ since there are 6 ways to get doubles and let B = getting a sum of 8. Then $P(B) = \frac{5}{36}$ since there are 5 ways to get a sum of 8—(6, 2), (5, 3), (4, 4), (3, 5), and (2, 6). Let (A and B) = the number of ways to get a double and a sum of 8. There is only one way for this event to occur—(4, 4); then $P(A \text{ and } B) = \frac{1}{36}$. Hence,

$$P(A \text{ or } B) = P(A) + P(B) - P(A \text{ and } B) = \frac{6}{36} + \frac{5}{36} - \frac{1}{36} = \frac{10}{36} = \frac{5}{18}$$

EXAMPLE: At a political rally, there are 8 Democrats and 10 Republicans. Six of the Democrats are females and 5 of the Republicans are females. If a person is selected at random, find the probability that the person is a female or a Democrat.

SOLUTION:

There are 18 people at the rally. Let $P(\text{female}) = \frac{6+5}{18} = \frac{11}{18}$ since there are 11 females, and $P(\text{Democrat}) = \frac{8}{18}$ since there are 8 Democrats. $P(\text{female and Democrat}) = \frac{6}{18}$ since 6 of the Democrats are females. Hence,

$$P(\text{female or Democrat}) = P(\text{female}) + P(\text{Democrat}) - P(\text{female and Democrat}) = \frac{11}{18} + \frac{8}{18} - \frac{6}{18} = \frac{13}{18}$$

EXAMPLE: The probability that a student owns a computer is 0.92, and the probability that a student owns an automobile is 0.53. If the probability that a student owns both a computer and an automobile is 0.49, find the probability that the student owns a computer or an automobile.

SOLUTION:

Since $P(\text{computer}) = 0.92$, $P(\text{automobile}) = 0.53$, and $P(\text{computer and automobile}) = 0.49$, $P(\text{computer or automobile}) = 0.92 + 0.53 - 0.49 = 0.96$.

The key word for addition is "or," and it means that one event or the other occurs. If the events are not mutually exclusive, the probability of the outcomes that the two events have in common must be subtracted from the sum of the probabilities of the two events. For the mathematical purist, only one addition rule is necessary, and that is

$$P(A \text{ or } B) = P(A) + P(B) - P(A \text{ and } B)$$

The reason is that when the events are mutually exclusive, $P(A \text{ and } B)$ is equal to zero because mutually exclusive events have no outcomes in common.

PRACTICE

1. When a card is selected at random from a 52-card deck, find the probability that the card is a face card or a spade.
2. A die is rolled. Find the probability that the result is an even number or a number less than 3.
3. Two dice are rolled. Find the probability that a number on one die is a six or the sum of the spots is eight.
4. A coin is tossed and a die is rolled. Find the probability that the coin falls heads up or that there is a 4 on the die.
5. In a psychology class, there are 15 sophomores and 18 juniors. Six of the sophomores are males and 10 of the juniors are males. If a student is selected at random, find the probability that the student is
   a. A junior or a male.
   b. A sophomore or a female.
   c. A junior.

ANSWERS

1. $P(\text{face card or spade}) = P(\text{face card}) + P(\text{spade}) - P(\text{face card and spade}) = \frac{12}{52} + \frac{13}{52} - \frac{3}{52} = \frac{22}{52} = \frac{11}{26}$
2. $P(\text{even or less than three}) = P(\text{even}) + P(\text{less than three}) - P(\text{even and less than three}) = \frac{3}{6} + \frac{2}{6} - \frac{1}{6} = \frac{4}{6} = \frac{2}{3}$
3. $P(6 \text{ or a sum of } 8) = P(6) + P(\text{sum of } 8) - P(6 \text{ and sum of } 8) = \frac{11}{36} + \frac{5}{36} - \frac{2}{36} = \frac{14}{36} = \frac{7}{18}$
4. $P(\text{heads or } 4) = P(\text{heads}) + P(4) - P(\text{heads and } 4) = \frac{1}{2} + \frac{1}{6} - \frac{1}{12} = \frac{7}{12}$
5. a. $P(\text{junior or male}) = P(\text{junior}) + P(\text{male}) - P(\text{junior and male}) = \frac{18}{33} + \frac{16}{33} - \frac{10}{33} = \frac{24}{33} = \frac{8}{11}$

   b. $P(\text{sophomore or female}) = P(\text{sophomore}) + P(\text{female}) -$

   $P(\text{sophomore or female}) = \frac{15}{33} + \frac{17}{33} - \frac{9}{33} = \frac{23}{33}$

   c. $P(\text{junior}) = \frac{18}{33} = \frac{6}{11}$

Summary

Many times in probability, it is necessary to find the probability of two or more events occurring. In these cases, the addition rules are used. When the events are mutually exclusive, addition rule I is used, and when the events are not mutually exclusive, addition rule II is used. If the events are mutually exclusive, they have no outcomes in common. When the two events are not mutually exclusive, they have some common outcomes. The key word in these problems is "or," and it means to add.

CHAPTER QUIZ

1. Which of the two events are not mutually exclusive?

   a. Rolling a die and getting a 6 or a 3
   b. Drawing a card from a deck and getting a club or an ace
   c. Tossing a coin and getting a head or a tail
   d. Tossing a coin and getting a head and rolling a die and getting an odd number

2. Which of the two events are mutually exclusive?

   a. Drawing a card from a deck and getting a king or a club
   b. Rolling a die and getting an even number or a 6
   c. Tossing two coins and getting two heads or two tails
   d. Rolling two dice and getting doubles or getting a sum of eight

3. In a box there are 6 white marbles, 3 blue marbles, and 1 red marble. If a marble is selected at random, what is the probability that it is red or blue?

   a. $\frac{2}{5}$

   b. $\frac{1}{3}$

   c. $\frac{9}{10}$

   d. $\frac{1}{9}$

4. When a single die is rolled, what is the probability of getting a prime number (2, 3, or 5)?

   a. $\frac{5}{6}$

   b. $\frac{2}{3}$

   c. $\frac{1}{2}$

   d. $\frac{1}{6}$

5. A storeowner plans to have his annual "Going Out of Business Sale." If each month has an equal chance of being selected, find the probability that the sale will be in a month that begins with the letter J or M.

   a. $\frac{1}{4}$

   b. $\frac{1}{6}$

   c. $\frac{5}{8}$

   d. $\frac{5}{12}$

6. A card is selected from a deck of 52 cards. Find the probability that it is a red queen or a black ace.

   a. $\frac{2}{13}$

   b. $\frac{1}{13}$

   c. $\frac{5}{13}$

   d. $\frac{8}{13}$

7. At a high school with 300 students, 62 play football, 33 play baseball, and 14 play both sports. If a student is selected at random, find the probability that the student plays football or baseball.

   a. $\frac{27}{100}$

   b. $\frac{109}{300}$

   c. $\frac{19}{60}$

   d. $\frac{14}{300}$

8. A card is selected from a deck. Find the probability that it is a face card or a diamond.

   a. $\frac{25}{52}$

   b. $\frac{3}{52}$

   c. $\frac{11}{26}$

   d. $\frac{13}{52}$

9. A single card is selected from a deck. Find the probability that it is a queen or a black card.

   a. $\frac{11}{26}$

   b. $\frac{7}{13}$

   c. $\frac{1}{13}$

   d. $\frac{15}{26}$

10. Two dice are rolled. What is the probability of getting doubles or a sum of 10?

    a. $\frac{11}{18}$

    b. $\frac{2}{9}$

    c. $\frac{1}{4}$

    d. $\frac{11}{36}$

11. The probability that a family visits Safari Zoo is 0.65, and the probability that a family rides on the Mt. Pleasant Tourist Railroad is 0.55. The probability that a family does both is 0.43. Find the probability that the family visits the zoo or the railroad.

    a. 0.77

    b. 0.22

    c. 0.12

    d. 0.10

12. If a card is drawn from a deck, what is the probability that it is a king, queen, or an ace?

    a. $\frac{5}{13}$

    b. $\frac{7}{13}$

    c. $\frac{6}{13}$

    d. $\frac{3}{13}$

Probability Sidelight

WIN A MILLION OR BE STRUCK BY LIGHTNING?

Do you think you are more likely to win a large lottery and become a millionaire or are you more likely to be struck by lightning?

Consider each probability. In a recent article, researchers estimated that the chance of winning a million or more dollars in a lottery is about one in 2 million. In a recent Pennsylvania State Lottery, the chances of winning a million dollars were 1 in 9.6 million. The chances of winning a $10 million prize in Publisher's Clearinghouse Sweepstakes were 1 in 2 million. Now the chances of being struck by lightning are about 1 in 600,000. Thus, a person is at least three times more likely to be struck by lightning than win a million dollars!

But wait a minute! Statisticians are critical of these types of comparisons, since winning the lottery is a random occurrence. But being struck by lightning depends on several factors. For example, if a person lives in a region where there are a lot of thunderstorms, his or her chances of being struck increase. Also, where a person is during a thunder storm influences his or her chances of being struck by lightning. If the person is in a safe place such as inside a building or in an automobile, the probability of being struck is relatively small compared to a person standing out in a field or on a golf course during a thunderstorm.

So be wary of such comparisons. As the old saying goes, you cannot compare apples and oranges.

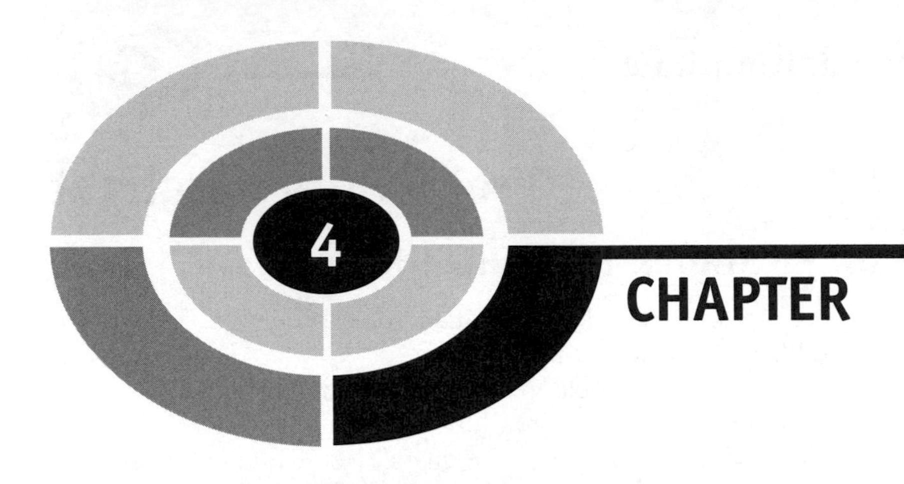

The Multiplication Rules

Introduction

The previous chapter showed how the addition rules could be used to solve problems in probability. This chapter will show you how to use the multiplication rules to solve many problems in probability. In addition, the concept of independent and dependent events will be introduced.

Independent and Dependent Events

The multiplication rules can be used to find the probability of two or more events that occur in sequence. For example, we can find the probability of selecting three jacks from a deck of cards on three sequential draws. Before explaining the rules, it is necessary to differentiate between **independent** and **dependent** events.

Two events, A and B, are said to be **independent** if the fact that event A occurs does not affect the probability that event B occurs. For example, if a coin is tossed and then a die is rolled, the outcome of the coin in no way affects or changes the probability of the outcome of the die. Another example would be selecting a card from a deck, replacing it, and then selecting a second card from a deck. The outcome of the first card, as long as it is replaced, has no effect on the probability of the outcome of the second card.

On the other hand, when the occurrence of the first event in some way changes the probability of the occurrence of the second event, the two events are said to be **dependent.** For example, suppose a card is selected from a deck and **not** replaced, and a second card is selected. In this case, the probability of selecting any specific card on the first draw is $\frac{1}{52}$, but since this card is not replaced, the probability of selecting any other specific card on the second draw is $\frac{1}{51}$, since there are only 51 cards left.

Another example would be parking in a no parking zone and getting a parking ticket. Again, if you are legally parked, the chances of getting a parking ticket are pretty close to zero (as long as the meter does not run out). However, if you are illegally parked, your chances of getting a parking ticket dramatically increase.

PRACTICE

Determine whether the two events are independent or dependent.

1. Tossing a coin and selecting a card from a deck
2. Driving on ice and having an accident
3. Drawing a ball from an urn, not replacing it, and then drawing a second ball
4. Having a high I.Q. and having a large hat size
5. Tossing one coin and then tossing a second coin

ANSWERS

1. Independent. Tossing a coin has no effect on drawing a card.
2. Dependent. In most cases, driving on ice will increase the probability of having an accident.
3. Dependent. Since the first ball is not replaced before the second ball is selected, it will change the probability of a specific second ball being selected.
4. Independent. To the best of the author's knowledge, no studies have been done showing any relationship between hat size and I.Q.
5. Independent. The outcome of the first coin does not influence the outcome of the second coin.

Multiplication Rule I

Before explaining the first multiplication rule, consider the example of tossing two coins. The sample space is HH, HT, TH, TT. From classical probability theory, it can be determined that the probability of getting two heads is $\frac{1}{4}$, since there is only one way to get two heads and there are four outcomes in the sample space. However, there is another way to determine the probability of getting two heads. In this case, the probability of getting a head on the first toss is $\frac{1}{2}$, and the probability of getting a head on the second toss is also $\frac{1}{2}$. So the probability of getting two heads can be determined by multiplying $\frac{1}{2} \cdot \frac{1}{2} = \frac{1}{4}$ This example illustrates the first multiplication rule.

Multiplication Rule I: For two independent events A and B, $P(A \text{ and } B) = P(A) \cdot P(B)$.

In other words, when two independent events occur in sequence, the probability that both events will occur can be found by multiplying the probabilities of each individual event.

The word **and** is the key word and means that both events occur in sequence and to multiply.

EXAMPLE: A coin is tossed and a die is rolled. Find the probability of getting a tail on the coin and a 5 on the die.

SOLUTION:

Since $P(\text{tail}) = \frac{1}{2}$ and $P(5) = \frac{1}{6}$; $P(\text{tail and } 5) = P(\text{tail}) \cdot P(5) = \frac{1}{2} \cdot \frac{1}{6} = \frac{1}{12}$. Note that the events are independent.

The previous example can also be solved using classical probability. Recall that the sample space for tossing a coin and rolling a die is

H1, H2, H3, H4, H5, H6

T1, T2, T3, T4, T5, T6

Notice that there are 12 outcomes in the sample space and only one outcome is a tail and a 5; hence, $P(\text{tail and } 5) = \frac{1}{12}$.

EXAMPLE: An urn contains 2 red balls, 3 green balls, and 5 blue balls. A ball is selected at random and its color is noted. Then it is replaced and another ball is selected and its color is noted. Find the probability of each of these:

a. Selecting 2 blue balls
b. Selecting a blue ball and then a red ball
c. Selecting a green ball and then a blue ball

SOLUTION:

Since the first ball is being replaced before the second ball is selected, the events are independent.

a. There are 5 blue balls and a total of 10 balls; therefore, the probability of selecting two blue balls with replacement is

$$P(\text{blue and blue}) = P(\text{blue}) \cdot P(\text{blue})$$
$$= \frac{5}{10} \cdot \frac{5}{10}$$
$$= \frac{25}{100} = \frac{1}{4}$$

b. There are 5 blue balls and 2 red balls, so the probability of selecting a blue ball and then a red ball with replacement is

$$P(\text{blue and red}) = P(\text{blue}) \cdot P(\text{red})$$
$$= \frac{5}{10} \cdot \frac{2}{10}$$
$$= \frac{10}{100} = \frac{1}{10}$$

c. There are 3 green balls and 5 blue balls, so the probability of selecting a green ball and then a blue ball with replacement is

$$\begin{aligned} P(\text{green and blue}) &= P(\text{green}) \cdot P(\text{blue}) \\ &= \frac{3}{10} \cdot \frac{5}{10} \\ &= \frac{15}{100} = \frac{3}{20} \end{aligned}$$

The multiplication rule can be extended to 3 or more events that occur in sequence, as shown in the next example.

EXAMPLE: A die is tossed 3 times. Find the probability of getting three 6s.

SOLUTION:

When a die is tossed, the probability of getting a six is $\frac{1}{6}$; hence, the probability of getting three 6s is

$$\begin{aligned} P(6 \text{ and } 6 \text{ and } 6) &= P(6) \cdot P(6) \cdot P(6) \\ &= \frac{1}{6} \cdot \frac{1}{6} \cdot \frac{1}{6} \\ &= \frac{1}{216} \end{aligned}$$

Another situation occurs in probability when subjects are selected from a large population. Even though the subjects are not replaced, the probability changes only slightly, so the change can be ignored. Consider the next example.

EXAMPLE: It is known that 66% of the students at a large college favor building a new fitness center. If two students are selected at random, find the probability that all of them favor the building of a new fitness center.

SOLUTION:

Since the student population at the college is large, selecting a student does not change the 66% probability that the next student selected will favor the building of a new fitness center; hence, the probability of selecting two students who both favor the building of a new fitness center is $(0.66)(0.66) = 0.4356$ or 43.56%.

PRACTICE

1. A card is drawn from a deck, then replaced, and a second card is drawn. Find the probability that two kings are selected.
2. If 12% of adults are left-handed, find the probability that if three adults are selected at random, all three will be left-handed.
3. If two people are selected at random, find the probability that both were born in August.
4. A coin is tossed 4 times. Find the probability of getting 4 heads.
5. A die is rolled and a card is selected at random from a deck of 52 cards. Find the probability of getting an odd number on the die and a club on the card.

ANSWERS

1. The probability that 2 kings are selected is

$$P(\text{king and king}) = P(\text{king}) \cdot P(\text{king}) = \frac{\not{4}^{1}}{\not{52}^{13}} \cdot \frac{\not{4}^{1}}{\not{52}^{13}} = \frac{1}{169}$$

2. The probability of selecting 3 adults who are left-handed is $(0.12)(0.12)(0.12) = 0.001728$.
3. Each person has approximately 1 chance in 12 of being born in August; hence, the probability that both are born in August is $\frac{1}{12} \cdot \frac{1}{12} = \frac{1}{144}$.
4. The probability of getting 4 heads is $\frac{1}{2} \cdot \frac{1}{2} \cdot \frac{1}{2} \cdot \frac{1}{2} = \frac{1}{16}$.
5. The probability of getting an odd number on the die is $\frac{3}{6} = \frac{1}{2}$, and the probability of getting a club is $\frac{13}{52} = \frac{1}{4}$; hence, the $P(\text{odd and club}) = P(\text{odd}) \cdot P(\text{club}) = \frac{1}{2} \cdot \frac{1}{4} = \frac{1}{8}$.

Multiplication Rule II

When two sequential events are dependent, a slight variation of the multiplication rule is used to find the probability of both events occurring. For example, when a card is selected from an ordinary deck of 52 cards the probability of getting a specific card is $\frac{1}{52}$, but the probability of getting a specific card on the second draw is $\frac{1}{51}$ since 51 cards remain.

EXAMPLE: Two cards are selected from a deck and the first card is **not** replaced. Find the probability of getting two kings.

SOLUTION:

The probability of getting a king on the first draw is $\frac{4}{52}$ and the probability of getting a king on the second draw is $\frac{3}{51}$, since there are 3 kings left and 51 cards left. Hence the probability of getting 2 kings when the first card is not replaced is $\frac{4}{52} \cdot \frac{3}{51} = \frac{12}{2652} = \frac{1}{221}$.

When the two events A and B are dependent, the probability that the second event B occurs after the first event A has already occurred is written as $P(B \mid A)$. This does not mean that B is divided by A; rather, it means and is read as "the probability that event B occurs *given* that event A has already occurred." $P(B \mid A)$ also means the **conditional** probability that event B occurs given event A has occurred. The second multiplication rule follows.

Multiplication Rule II: When two events are dependent, the probability of both events occurring is $P(A \text{ and } B) = P(A) \cdot P(B \mid A)$

EXAMPLE: A box contains 24 toasters, 3 of which are defective. If two toasters are selected and tested, find the probability that both are defective.

SOLUTION:

Since there are 3 defective toasters out of 24, the probability that the first toaster is defective is $\frac{3}{24} = \frac{1}{8}$. Since the second toaster is selected from the remaining 23 and there are two defective toasters left, the probability that it is defective is $\frac{2}{23}$. Hence, the probability that both toasters are defective is

$$P(D_1 \text{ and } D_2) = P(D_1) \cdot P(D_2|D_1) = \frac{\cancel{3}^1}{\cancel{24}\,\cancel{8}^4} \cdot \frac{\cancel{2}^1}{23} = \frac{1}{92}$$

EXAMPLE: Two cards are drawn without replacement from a deck of 52 cards. Find the probability that both are queens.

SOLUTION:

$$\begin{aligned} P(Q \text{ and } Q) &= P(Q) \cdot P(Q|Q) \\ &= \frac{4}{52} \cdot \frac{3}{51} \\ &= \frac{1}{221} \end{aligned}$$

This multiplication rule can be extended to include three or more events, as shown in the next example.

EXAMPLE: A box contains 3 orange balls, 3 yellow balls, and 2 white balls. Three balls are selected without replacement. Find the probability of selecting 2 yellow balls and a white ball.

SOLUTION:

$$\begin{aligned} P(\text{yellow and yellow and white}) &= \frac{3}{8} \cdot \frac{2}{7} \cdot \frac{2}{6} \\ &= \frac{\cancel{3}^1}{\cancel{8}^4} \cdot \frac{\cancel{2}^1}{7} \cdot \frac{\cancel{2}^1}{\cancel{6}^1} \\ &= \frac{1}{28} \end{aligned}$$

Remember that the key word for the multiplication rule is *and*. It means to multiply.

When two events are dependent, the probability that the second event occurs must be adjusted for the occurrence of the first event. For the mathematical purist, only one multiplication rule is necessary for two events, and that is

$$P(A \text{ and } B) = P(A) \cdot P(B \mid A).$$

The reason is that when the events are independent $P(B|A) = P(B)$ since the occurrence of the first event A has no effect on the occurrence of the second event B.

PRACTICE

1. In a study, there are 8 guinea pigs; 5 are black and 3 are white. If 2 pigs are selected without replacement, find the probability that both are white.
2. In a classroom there are 8 freshmen and 6 sophomores. If three students are selected at random for a class project, find the probability that all 3 are freshmen.
3. Three cards are drawn from a deck of 52 cards without replacement. Find the probability of getting 3 diamonds.
4. A box contains 12 calculators of which 5 are defective. If two calculators are selected without replacement, find the probability that both are good.
5. A large flashlight has 6 batteries. Three are dead. If two batteries are selected at random and tested, find the probability that both are dead.

ANSWERS

1. $P(\text{white and white}) = \frac{3}{\cancel{8}^4} \cdot \frac{\cancel{2}^1}{7} = \frac{3}{28}$
2. $P(\text{3 freshmen}) = \frac{8}{\cancel{14}^2} \cdot \frac{\cancel{7}^1}{13} \cdot \frac{\cancel{6}^1}{\cancel{12}^2} = \frac{2}{13}$
3. $P(\text{3 diamonds}) = \frac{\cancel{13}^1}{\cancel{52}^4} \cdot \frac{12}{51} \cdot \frac{11}{50} = \frac{11}{850}$
4. $P(\text{2 good}) = \frac{7}{\cancel{12}^2} \cdot \frac{\cancel{6}^1}{11} = \frac{7}{22}$
5. $P(\text{2 batteries dead}) = \frac{3}{\cancel{6}^3} \cdot \frac{\cancel{2}^1}{5} = \frac{3}{15} = \frac{1}{5}$

Conditional Probability

Previously, conditional probability was used to find the probability of sequential events occurring when they were dependent. Recall that $P(B|A)$ means the probability of event B occurring given that event A has already occurred. Another situation where conditional probability can be used is when additional information about an event is known. Sometimes it might be known that some outcomes in the sample space have occurred or that some outcomes cannot occur. When conditions are imposed or known on events, there is a possibility that the probability of the certain event occurring may change. For example, suppose you want to determine the probability that a house will be destroyed by a hurricane. If you used all houses in the United States as the sample space, the probability would be very small. However, if you used only the houses in the states that border the Atlantic Ocean as the sample space, the probability would be much higher. Consider the following examples.

EXAMPLE: A die is rolled; find the probability of getting a 4 if it is known that an even number occurred when the die was rolled.

SOLUTION:

If it is known that an even number has occurred, the sample space is reduced to 2, 4, or 6. Hence the probability of getting a 4 is $\frac{1}{3}$ since there is one chance in three of getting a 4 if it is known that the result was an even number.

EXAMPLE: Two dice are rolled. Find the probability of getting a sum of 3 if it is known that the sum of the spots on the dice was less than six.

SOLUTION:

There are 2 ways to get a sum of 3. They are (1, 2) and (2, 1), and there are 10 ways to get a sum less than six. They are (1, 1), (1, 2), (2, 1), (3, 1), (2, 2), (1, 3), (1, 4), (2, 3), (3, 2), and (4, 1); hence, $P(\text{sum of } 3|\text{sum less than } 6) = \frac{2}{10} = \frac{1}{5}$.

The two previous examples of conditional probability were solved using classical probability and reduced sample spaces; however, they can be solved by using the following formula for conditional probability.

The conditional probability of two events A and B is

$$P(A|B) = \frac{P(A \text{ and } B)}{P(B)}.$$

$P(A \text{ and } B)$ means the probability of the outcomes that events A and B have in common. The two previous examples will now be solved using the formula for conditional probability.

EXAMPLE: A die is rolled; find the probability of getting a 4, if it is known that an even number occurred when the die was rolled.

SOLUTION:

$P(A \text{ and } B)$ is the probability of getting a 4 and an even number at the same time. Notice that there is only one way to get a 4 and an even number—the outcome 4. Hence $P(A \text{ and } B) = \frac{1}{6}$. Also $P(B)$ is the probability of getting an even number which is $\frac{3}{6} = \frac{1}{2}$. Now

$$P(A|B) = \frac{P(A \text{ and } B)}{P(B)} = \frac{\frac{1}{6}}{\frac{1}{2}} = \frac{1}{6} \div \frac{1}{2} = \frac{1}{\cancel{6}^{3}} \cdot \frac{\cancel{2}^{1}}{1} = \frac{1}{3}$$

Notice that the answer is the same as the answer obtained when classical probability was used.

EXAMPLE: Two dice are rolled. Find the probability of getting a sum of 3 if it is known that the sum of the spots on the dice was less than 6.

SOLUTION:

$P(A \text{ and } B)$ means the probability of getting a sum of 3 and a sum less than 6. Hence $P(A \text{ and } B) = \frac{2}{36}$. $P(B)$ means getting a sum less than 6 and is $\frac{10}{36}$. Hence,

$$P(A|B) = \frac{P(A \text{ and } B)}{P(B)} = \frac{\frac{2}{36}}{\frac{10}{36}} = \frac{2}{36} \div \frac{10}{36} = \frac{\cancel{2}^{1}}{\cancel{36}} \cdot \frac{\cancel{36}}{\cancel{10}^{5}} = \frac{1}{5}$$

EXAMPLE: When two dice were rolled, it is known that the sum was an even number. In this case, find the probability that the sum was 8.

SOLUTION:

When rolling two dice, there are 18 outcomes in which the sum is an even number. There are 5 ways to get a sum of 8; hence, $P(\text{sum of } 8 | \text{sum is even})$ is $\frac{8}{18}$ or $\frac{4}{9}$.

EXAMPLE: In a large housing plan, 35% of the homes have a deck and a two-car garage, and 80% of the houses have a two-car garage. Find the probability that a house has a deck given that it has a garage.

SOLUTION:

$$P(\text{deck and two-car garage}) = 0.35 \text{ and } P(\text{two-car garage}) = 0.80\text{; hence,}$$

$$P(\text{deck}|\text{two-car garage}) = \frac{(\text{deck and two-car garage})}{P(\text{two-car garage})} = \frac{0.35}{0.80} = \frac{7}{16} \text{ or } 0.4375$$

PRACTICE

1. When two dice are rolled, find the probability that one die is a 6 given that the sum of the spots is 8.
2. Two coins are tossed. Find the probability of getting two tails if it is known that one of the coins is a tail.
3. A card is selected from a deck. Find the probability that it is an ace given that it is a black card.
4. The probability that a family visits the Sand Crab Water Park and the Rainbow Gardens Amusement Park is 0.20. The probability that a family visits the Rainbow Gardens Amusement Park is 0.80. Find the probability that a family visits the Sand Crab Water Park if it is known that they have already visited Rainbow Gardens Amusement Park.
5. Three dice are rolled. Find the probability of getting three twos if it is known that the sum of the spots of the three dice was six.

ANSWERS

1. There are five ways to get a sum of 8. They are (6, 2), (2, 6), (3, 5), (5, 3), and (4, 4). There are two ways to get a sum of 8 when one die is a 6. Hence,

$$P(\text{one die is a six}|\text{sum of 8}) = \frac{P(\text{one die is a six and sum of 8})}{P(\text{sum of 8})}$$

$$= \frac{\frac{2}{36}}{\frac{5}{36}}$$

$$= \frac{2}{36} \div \frac{5}{36} = \frac{2}{\cancel{36}^1} \cdot \frac{\cancel{36}^1}{5}$$

$$= \frac{2}{5}$$

The problem can also be solved by looking at the reduced sample space. There are two possible ways one die is a 6 and the sum of the dice is 8. There are five ways to get a sum of 8. Hence, the probability is $\frac{2}{5}$.

2. There are 3 ways to get one tail: HT, TH, and TT. There is one way to get two tails; hence, the probability of getting two tails given that one of the coins is a tail is $\frac{1}{3}$. The problem can also be solved using the formula for conditional probability:

$$P(\text{two tails}|\text{given one coin was a tail}) = \frac{\frac{1}{4}}{\frac{3}{4}}$$

$$= \frac{1}{4} \div \frac{3}{4}$$

$$= \frac{1}{\cancel{4}^1} \div \frac{\cancel{4}^1}{3}$$

$$= \frac{1}{3}$$

3. There are 2 black aces and 26 black cards; hence, $P(\text{ace}|\text{black card}) = \frac{2}{26}$ or $\frac{1}{13}$ or

$$P(\text{ace}|\text{black card}) = \frac{P(\text{ace and black card})}{P(\text{black card})} = \frac{\frac{2}{52}}{\frac{26}{52}} = \frac{2}{52} \div \frac{26}{52} = \frac{\cancel{2}^1}{\cancel{52}^1} \cdot \frac{\cancel{52}^1}{\cancel{26}^{13}} = \frac{1}{13}$$

4. P (Sand Crab|Rainbow Gardens)

$$= \frac{P(\text{Sand Crab and Rainbow Gardens})}{P(\text{Rainbow Gardens})} = \frac{0.20}{0.80} = 0.25$$

5. There is one way to get 3 twos—(2, 2, 2) and there are 10 ways to get a sum of six—(2, 2, 2), (3, 2, 1), (1, 2, 3), (2, 1, 3), (2, 3, 1), (3, 1, 2), (1, 3, 2), (4, 1, 1), (1, 4, 1,) and (1, 1, 4). Hence,

$$P(\text{3 twos}|\text{sum of 6}) = \frac{P(\text{3 twos and sum of 6})}{P(\text{sum of 6})} = \frac{\frac{1}{36}}{\frac{10}{36}} = \frac{1}{36} \div \frac{10}{36} = \frac{1}{\cancel{36}^1} \cdot \frac{\cancel{36}^1}{10} = \frac{1}{10}$$

Summary

When two events occur in sequence, the probability that both events occur can be found by using one of the multiplication rules. When two events are independent, the probability that the first event occurs does not affect or change the probability of the second event occurring. If the events are independent, multiplication rule I is used. When the two events are dependent, the probability of the second event occurring is changed after the first event occurs. If the events are dependent, multiplication rule II is used. The key word for using the multiplication rule is "and." Conditional probability is used when additional information is known about the probability of an event.

CHAPTER QUIZ

1. Which of the following events are dependent?

   a. Tossing a coin and rolling a die
   b. Rolling a die and then rolling a second die
   c. Sitting in the sun all day and getting sunburned
   d. Drawing a card from a deck and rolling a die

2. Three dice are rolled. What is the probability of getting three 4s?

   a. $\frac{1}{216}$

   b. $\frac{1}{6}$

   c. $\frac{1}{36}$

   d. $\frac{1}{18}$

3. What is the probability of selecting 4 spades from a deck of 52 cards if each card is replaced before the next one is selected?

   a. $\frac{4}{13}$

   b. $\frac{1}{2561}$

   c. $\frac{1}{52}$

   d. $\frac{1}{13}$

4. A die is rolled five times. What is the probability of getting 5 twos?

   a. $\frac{1}{8}$

   b. $\frac{1}{4}$

   c. $\frac{1}{6}$

   d. $\frac{1}{32}$

5. A coin is tossed four times; what is the probability of getting 4 heads?

   a. $\frac{1}{2}$

   b. $\frac{1}{16}$

   c. 1

   d. $\frac{1}{4}$

6. If 25% of U.S. prison inmates are not U.S. citizens, what is the probability of randomly selecting three inmates who will not be U.S. citizens?

   a. 0.75
   b. 0.421875
   c. 0.015625
   d. 0.225

7. If three people are randomly selected, find the probability that they will all have birthdays in June.

   a. $\frac{1}{1728}$
   b. $\frac{1}{36}$
   c. $\frac{1}{12}$
   d. $\frac{1}{4}$

8. In a sample of 10 telephones, 4 are defective. If 3 are selected at random and tested, what is the probability that all will be non-defective?

   a. $\frac{1}{30}$
   b. $\frac{8}{125}$
   c. $\frac{1}{6}$
   d. $\frac{27}{125}$

9. A bag contains 4 blue marbles and 5 red marbles. If 2 marbles are selected at random without replacement, what is the probability that both will be blue?

   a. $\frac{16}{81}$
   b. $\frac{1}{6}$
   c. $\frac{1}{2}$
   d. $\frac{1}{15}$

10. If two people are selected at random from the phone book of a large city, find the probability that they were both born on a Sunday.

    a. $\frac{1}{7}$

    b. $\frac{1}{49}$

    c. $\frac{2}{365}$

    d. $\frac{2}{7}$

11. The numbers 1 to 12 are placed in a hat, and a number is selected. What is the probability that the number is 4 given that it is known to be an even number?

    a. $\frac{2}{3}$

    b. $\frac{3}{4}$

    c. $\frac{1}{6}$

    d. $\frac{1}{8}$

12. Three coins are tossed; what is the probability of getting 3 heads if it is known that at least two heads were obtained?

    a. $\frac{1}{4}$

    b. $\frac{2}{3}$

    c. $\frac{1}{2}$

    d. $\frac{3}{8}$

13. In a certain group of people, it is known that 40% of the people take Vitamins C and E on a daily basis. It is known that 75% take Vitamin C on a daily basis. If a person is selected at random, what is the probability that the person takes Vitamin E given that the person takes Vitamin C?

    a. $\frac{5}{13}$

    b. $\frac{8}{15}$

    c. $\frac{3}{11}$

    d. $\frac{7}{12}$

14. Two dice are tossed; what is the probability that the numbers are the same on both dice if it is known that the sum of the spots is 6?

    a. $\frac{2}{3}$

    b. $\frac{1}{6}$

    c. $\frac{4}{5}$

    d. $\frac{1}{5}$

Probability Sidelight

THE LAW OF AVERAGES

Suppose I asked you that if you tossed a coin 9 times and got 9 heads, what would you bet that you would get on the tenth toss, heads or tails? Most people would bet on a tail. When asked why they would select a tail, they would probably respond that a tail was "due" according to the "law of averages." In reality, however, the probability of getting a head on the tenth toss is $\frac{1}{2}$, and the probability of getting a tail on the tenth toss is $\frac{1}{2}$, so it doesn't really matter since the probabilities are the same. A coin is an inanimate object. It does not have a memory. It doesn't know that in the long run, the number of heads and the number of tails should balance out. So does that make the law of averages wrong? No. You see, there's a big difference between asking the question, "What is the probability of getting 10 heads if I toss a coin ten times?" and "If I get 9 heads in a row, what is the probability of getting a head on the tenth toss?" The answer to the first question is $\frac{1}{2^{10}} = \frac{1}{1024}$, that is about 1 chance in 1000, and the answer to the second question is $\frac{1}{2}$.

This reasoning can be applied to many situations. For example, suppose that a prize is offered for tossing a coin and getting 10 heads in a row. If you played the game, you would have only one chance in 1024 of winning, but if 1024 people played the game, there is a pretty good chance that somebody would win the prize. If 2028 people played the game, there would be a good chance that two people might win. So what does this mean? It means that the probability of winning big in a lottery or on a slot machine is very small, but since there are many, many people playing, somebody will probably win; however, your chances of winning big are very small.

A similar situation occurs when couples have children. Suppose a husband and wife have four boys and would like to have a girl. It is incorrect to reason that the chance of having a family of 5 boys is $\frac{1}{32}$, so it is more likely that the next child will be a girl. However, after each child is born, the probability that the next child is a girl (or a boy for that matter) is about $\frac{1}{2}$. The law of averages is not appropriate here.

My wife's aunt had seven girls before the first boy was born. Also, in the Life Science Library's book entitled *Mathematics,* there is a photograph of the Landon family of Harrison, Tennessee, that shows Mr. and Mrs. Emery Landon and their 13 boys!

Another area where people incorrectly apply the law of averages is in attempting to apply a betting system to gambling games. One such system is doubling your bet when you lose. Consider a game where a coin is tossed. If it lands heads, you win what you bet. If it lands tails, you lose. Now if you bet one dollar on the first toss and get a head, you win one dollar. If you get tails, you lose one dollar and bet two dollars on the next toss. If you win, you are one dollar ahead because you lost one dollar on the first bet but won two dollars on the second bet. If you get a tail on the second toss, you bet four dollars on the third toss. If you win, you start over with a one dollar bet, but if you lose, you bet eight dollars on the next toss. With this system, you win every time you get a head. Sounds pretty good, doesn't it?

This strategy won't work because if you play long enough, you will eventually run out of money since if you get a series of tails, you must increase your bet substantially each time. So if you lose five times in a row, you have lost \$1 + \$2 + \$4 + \$8 + \$16 or \$31, and your next bet has to be \$32. So you are betting \$63 to win \$1. Runs do occur and when they do, hope that they are in your favor.

Now let's look at some unusual so-called "runs."

In 1950, a person won 28 straight times playing the game of craps (dice) at the Desert Inn in Las Vegas. He lost on the twenty-ninth roll. He did not win big though because after each win he stuffed some bills in his pocket. The event took about one hour and twenty minutes.

In 1959 in a casino in Puerto Rico at a roulette game, the number 10 occurred six times in succession. There are 38 numbers on a roulette wheel.

At a casino in New York in 1943 the color red occurred in a roulette game 32 times in a row, and at a casino in Monte Carlo an even number occurred in a roulette game 28 times in a row.

These incidents have been reported in two books, one entitled *Scarne's Complete Guide to Gambling* and the other entitled *Lady Luck* by Warren Weaver.

So what can be concluded? First, rare events (events with a small probability of occurring) can and do occur. Second, the more people who play a game, the more likely someone will win. Finally, the law of averages applies when there is a large number of independent outcomes in which the probability of each outcome occurring does not change.

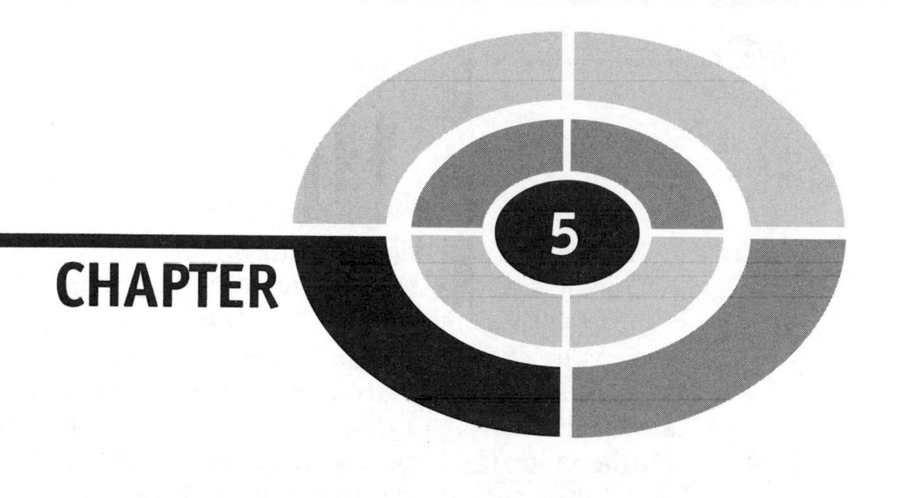

Odds and Expectation

Introduction

In this chapter you will learn about two concepts that are often used in conjunction with probability. They are **odds** and **expectation.** Odds are used most often in gambling games at casinos and racetracks, and in sports betting and lotteries. Odds make it easier than probabilities to determine payoffs.

Mathematical expectation can be thought of more or less as an average over the long run. In other words, if you would perform a probability experiment many times, the expectation would be an average of the outcomes. Also, expectation can be used to determine the average payoff per game in a gambling game.

Odds

Odds are used by casinos, racetracks, and other gambling establishments to determine payoffs when bets are made. For example, at a race, the odds that a horse wins the race may be 4 to 1. In this case, if you bet \$1 and the horse wins, you get \$4. If you bet \$2 and the horse wins, you get \$8, and so on.

Odds are computed from probabilities. For example, suppose you roll a die and if you roll a three, you win. If you roll any other number, you lose. Furthermore, if you bet one dollar and win, what would the payoff be if you win? In this case, there are six outcomes, and you have one chance (outcome) of winning, so the probability that you win is $\frac{1}{6}$. That means on *average* you win once in every six rolls. So if you lose on the first five rolls and win on the sixth, you have lost \$5 and therefore, you should get \$5 if you win on the sixth roll. So if you bet \$1 and win \$5, the odds are 1 to 5. Of course, there is no guarantee that you will win on the sixth roll. You may win on the first roll or any roll, but on *average* for every six rolls, you will win one time over the long run.

In gambling games, the odds are expressed backwards. For example, if there is one chance in six that you will win, the odds are 1 to 5, but in general, the odds would be given as 5 to 1. In gambling, the house (the people running the game) will offer lower odds, say 4 to 1, in order to make a profit. In this case, then, the player wins on average one time in every 6 rolls and spends on average \$5, but when the player wins, he gets only \$4. So the house wins on average \$1 for every six rolls of the player.

Odds can be expressed as a fraction, $\frac{1}{5}$, or as a ratio, $1:5$. If the odds of winning the game are $1:5$, then the odds of losing are $5:1$. The odds of winning the game can also be called the odds "in favor" of the event occurring. The odds of losing can also be called "the odds against" the event occurring.

The formulas for odds are

$$\text{odds in favor} = \frac{P(E)}{1 - P(E)}$$

$$\text{odds against} = \frac{P(\overline{E})}{1 - P(\overline{E})}$$

where $P(E)$ is the probability that the event E occurs and $P(\overline{E})$ is the probability that the event does not occur.

EXAMPLE: Two coins are tossed; find the odds in favor of getting two heads

SOLUTION:

When two coins are tossed, there are four outcomes and $P(\text{HH}) = \frac{1}{4}$. $P(\overline{E}) = 1 - \frac{1}{4} = \frac{3}{4}$; hence,

$$\text{odds in favor of two heads} = \frac{P(E)}{1 - P(E)} = \frac{\frac{1}{4}}{1 - \frac{1}{4}} = \frac{\frac{1}{4}}{\frac{3}{4}}$$

$$= \frac{1}{4} \div \frac{3}{4} = \frac{1}{\cancel{4}^1} \cdot \frac{\cancel{4}^1}{3} = \frac{1}{3}$$

The odds are 1 : 3.

EXAMPLE: Two dice are rolled; find the odds against getting a sum of 9.

SOLUTION:

There are 36 outcomes in the sample space and four ways to get a sum of 9. $P(\text{sum of } 9) = \frac{4}{36} = \frac{1}{9}$, $P(\overline{E}) = 1 - \frac{1}{9} = \frac{8}{9}$. Hence,

$$\text{odds of not getting a sum of } 9 = \frac{P(\overline{E})}{1 - P(\overline{E})} = \frac{\frac{8}{9}}{1 - \frac{8}{9}} = \frac{\frac{8}{9}}{\frac{1}{9}}$$

$$= \frac{8}{9} \div \frac{1}{9} = \frac{8}{\cancel{9}^1} \cdot \frac{\cancel{9}^1}{1} = \frac{8}{1}$$

The odds are 8 : 1.

If the odds in favor of an event occurring are $A : B$, then the odds against the event occurring are $B : A$. For example, if the odds are 1 : 15 that an event will occur, then the odds against the event occurring are 15 : 1.

Odds can also be expressed as

$$\text{odds in favor} = \frac{\text{number of outcomes in favor of the event}}{\text{number of outcomes not in favor of the event}}$$

For example, if two coins are tossed, the odds in favor of getting two heads were computed previously as 1 : 3. Notice that there is only one way to get two heads (HH) and three ways of not getting two heads (HT, TH, TT); hence the odds are 1 : 3.

When the probability of an event occurring is $\frac{1}{2}$, then the odds are 1 : 1. In the realm of gambling, we say the odds are "even" and the chance of the event is "fifty–fifty." The game is said to be fair. Odds can be other numbers, such as 2 : 5, 7 : 4, etc.

PRACTICE

1. When two dice are rolled, find the odds in favor of getting a sum of 12.
2. When a single card is drawn from a deck of 52 cards, find the odds against getting a diamond.
3. When three coins are tossed, find the odds in favor of getting two tails and a head in any order.
4. When a single die is rolled, find the odds in favor of getting an even number.
5. When two dice are rolled, find the odds against getting a sum of 7.

ANSWERS

1. There is only one way to get a sum of 12, and that is (6, 6). There are 36 outcomes in the sample space. Hence, $P(\text{sum of } 12) = \frac{1}{36}$. The odds in favor are

$$\frac{\frac{1}{36}}{1-\frac{1}{36}} = \frac{\frac{1}{36}}{\frac{35}{36}} = \frac{1}{36} \div \frac{35}{36} = \frac{1}{\cancel{36}^1} \cdot \frac{\cancel{36}^1}{35} = \frac{1}{35}$$

The odds are 1 : 35.

2. There are 13 diamonds in 52 cards; hence, $P(\blacklozenge) = \frac{13}{52} = \frac{1}{4}$. $P(\overline{\blacklozenge}) = 1 - \frac{1}{4} = \frac{3}{4}$. The odds against getting a diamond

$$\frac{\frac{3}{4}}{1-\frac{3}{4}} = \frac{\frac{3}{4}}{\frac{1}{4}} = \frac{3}{4} \div \frac{1}{4} = \frac{3}{\cancel{4}^1} \cdot \frac{\cancel{4}^1}{1} = \frac{3}{1}$$

The odds are 3 : 1.

3. When three coins are tossed, there are three ways to get two tails and a head. They are (TTH, THT, HTT), and there are eight outcomes in the sample space. The odds in favor of getting two tails and a head are

$$\frac{\frac{3}{8}}{1-\frac{3}{8}} = \frac{\frac{3}{8}}{\frac{5}{8}} = \frac{3}{8} \div \frac{5}{8} = \frac{3}{\cancel{8}^1} \cdot \frac{\cancel{8}^1}{5} = \frac{3}{5}$$

The odds are 3 : 5.

4. There are 3 even numbers out of 6 outcomes; hence, $P(\text{even}) = \frac{3}{6} = \frac{1}{2}$. The odds in favor of an even number are

$$\frac{\frac{1}{2}}{1-\frac{1}{2}} = \frac{\frac{1}{2}}{\frac{1}{2}} = \frac{1}{2} \div \frac{1}{2} = \frac{1}{\cancel{2}^1} \cdot \frac{\cancel{2}^1}{1} = \frac{1}{1}$$

The odds are 1:1.

5. There are six ways to get a sum of 7 and 36 outcomes in the sample space. Hence, $P(\text{sum of } 7) = \frac{6}{36} = \frac{1}{6}$ and $P(\text{not getting a sum of } 7) = 1 - \frac{1}{6} = \frac{5}{6}$. The odds against getting a sum of 7 are

$$\frac{\frac{5}{6}}{1-\frac{5}{6}} = \frac{\frac{5}{6}}{\frac{1}{6}} = \frac{5}{6} \div \frac{1}{6} = \frac{5}{\cancel{6}^1} \cdot \frac{\cancel{6}^1}{1} = \frac{5}{1}$$

The odds are 5 : 1.

Previously it was shown that given the probability of an event, the odds in favor of the event occurring or the odds against the event occurring can be found. The opposite is also true. If you know the odds in favor of an event occurring or the odds against an event occurring, you can find the probability of the event occurring. If the odds in favor of an event occurring are $A : B$, then the probability that the event will occur is $P(E) = \frac{A}{A+B}$.

If the odds against the event occurring are $B : A$, the probability that the event will not occur is $P(\overline{E}) = \frac{B}{B+A}$.

Note: Recall that $P(\overline{E})$ is the probability that the event will not occur or the probability of the complement of event E.

EXAMPLE: If the odds that an event will occur are 5 : 7, find the probability that the event will occur.

SOLUTION:

In this case, $A=5$ and $B=7$; hence, $P(E)=\frac{A}{A+B}=\frac{5}{5+7}=\frac{5}{12}$. Hence, the probability the event will occur is $\frac{5}{12}$.

EXAMPLE: If the odds in favor of an event are 2 : 9, find the probability that the event will **not** occur.

SOLUTION:

In this case, $A=2$ and $B=9$; hence, the probability that the event will not occur is

$$P(\overline{E})=\frac{B}{B+A}=\frac{9}{9+2}=\frac{9}{11}.$$

PRACTICE

1. Find the probability that an event E will occur if the odds are 5:2 in favor of E.
2. Find the probability that an event E will not occur if the odds against the event E are 4 : 1.
3. Find the probability that an event E will occur if the odds in favor of the event are 2 : 3.
4. When two dice are rolled, the odds in favor of getting a sum of 8 are 5 : 31; find the probability of getting a sum of 8.
5. When a single card is drawn from a deck of 52 cards, the odds against getting a face card are 10 : 3, find the probability of selecting a face card.

ANSWERS

1. Let $A=5$ and $B=2$; then $P(E)=\dfrac{5}{5+2}=\dfrac{5}{7}$.
2. Let $B=4$ and $A=1$; then $P(\overline{E})=\dfrac{4}{4+1}=\dfrac{4}{5}$.
3. Let $A=2$ and $B=3$; then $P(E)=\dfrac{2}{2+3}=\dfrac{2}{5}$.

4. Let $A=5$ and $B=31$; then $P(E)=\frac{5}{5+31}=\frac{5}{36}$.

5. Let $B=10$ and $A=3$ then $P(E)=\frac{3}{3+10}=\frac{3}{13}$.

Expectation

When a person plays a slot machine, sometimes the person wins and other times—most often—the person loses. The question is, "How much will the person win or lose in the long run?" In other words, what is the person's expected gain or loss? Although an individual's exact gain or exact loss cannot be computed, the overall gain or loss of all people playing the slot machine can be computed using the concept of mathematical expectation.

Expectation or **expected value** is a **long run average**. The expected value is also called the **mean,** and it is used in games of chance, insurance, and in other areas such as decision theory. The outcomes must be numerical in nature. The expected value of the outcome of a probability experiment can be found by multiplying each outcome by its corresponding probability and adding the results.

Formally defined, the expected value for the outcomes of a probability experiment is $E(X)=X_1 \cdot P(X_1)+X_2 \cdot P(X_2)+\cdots+X_n \cdot P(X_n)$ where the X corresponds to an outcome and the $P(X)$ to the corresponding probability of the outcome.

EXAMPLE: Find the expected value of the number of spots when a die is rolled.

SOLUTION:

There are 6 outcomes when a die is rolled. They are 1, 2, 3, 4, 5, and 6, and each outcome has a probability of $\frac{1}{6}$ of occurring, so the expected value of the numbered spots is $E(X)=1\cdot\frac{1}{6}+2\cdot\frac{1}{6}+3\cdot\frac{1}{6}+4\cdot\frac{1}{6}+5\cdot\frac{1}{6}+6\cdot\frac{1}{6}=\frac{21}{6}=3\frac{1}{2}$ or 3.5.

The expected value is 3.5.

Now what does this mean? When a die is rolled, it is not possible to get 3.5 spots, but if a die is rolled say 100 times and the average of the spots is computed, that average should be close to 3.5 if the die is fair. In other words, 3.5 is the theoretical or long run average. For example, if you rolled a die and were given \$1 for each spot on each roll, sometimes you would win \$1,

\$2, \$3, \$4, \$5, or \$6; however, on average, you would win \$3.50 on each roll. So if you rolled the die 100 times, you would win on average $\$3.50 \times 100 = \$350$. Now if you had to pay to play this game, you should pay \$3.50 for each roll. That would make the game fair. If you paid more to play the game, say \$4.00 each time you rolled the die, you would lose on average \$0.50 on each roll. If you paid \$3.00 to play the game, you would win an average \$0.50 per roll.

EXAMPLE: When two coins are tossed, find the expected value for the number of heads obtained.

SOLUTION:

Consider the sample space when two coins are tossed.

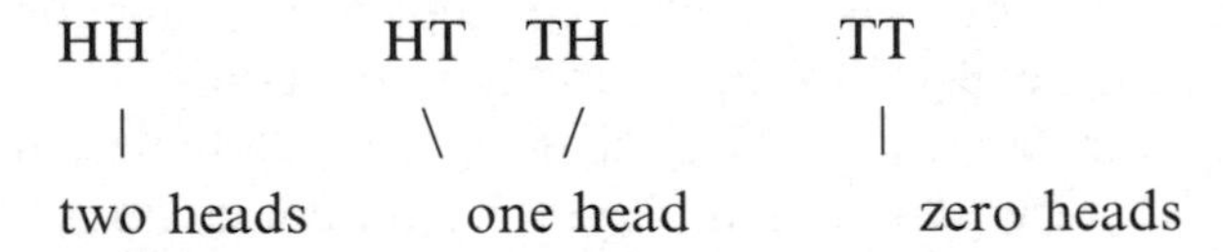

The probability of getting two heads is $\frac{1}{4}$. The probability of getting one head is $\frac{1}{4} + \frac{1}{4} = \frac{1}{2}$. The probability of getting no heads is $\frac{1}{4}$. The expected value for the number of heads is $E(X) = 2 \cdot \frac{1}{4} + 1 \cdot \frac{1}{2} + 0 \cdot \frac{1}{4} = 1$.

Hence the **average** number of heads obtained on each toss of 2 coins is 1.

In order to find the expected value for a gambling game, multiply the amount you win by the probability of winning that amount, and then multiply the amount you lose by the probability of losing that amount, then add the results. Winning amounts are positive and losses are negative.

EXAMPLE: One thousand raffle tickets are sold for a prize of an entertainment center valued at \$750. Find the expected value of the game if a person buys one ticket.

SOLUTION:

The problem can be set up as follows:

	Win	**Lose**
Gain, (X)	\$749	−\$1
Probability, $P(X)$	$\frac{1}{1000}$	$\frac{999}{1000}$

Since the person who buys a ticket does not get his or her \$1 back, the net gain if he or she wins is $\$750 - \$1 = \$749$. The probability of winning is one chance in 1000 since 1000 tickets are sold. The net loss is \$1 denoted as negative and the chances of not winning are $\frac{1000-1}{1000}$ or $\frac{999}{1000}$. Now $E(X) = \$749 \cdot \frac{1}{1000} + (-\$1)\frac{999}{1000} = -\$0.25$.

Here again it is necessary to realize that one cannot lose \$0.25 but what this means is that the house makes \$0.25 on every ticket sold. If a person purchased one ticket for raffles like this one over a long period of time, the person would lose **on average** \$0.25 each time since he or she would win on average one time in 1000.

There is an alternative method that can be used to solve problems when tickets are sold or when people pay to play a game. In this case, multiply the prize value by the probability of winning and subtract the cost of the ticket or the cost of playing the game. Using the information in the previous example, the solution looks like this:

$$E(X) = \$750 \cdot \frac{1}{1000} - \$1 = \$0.75 - \$1 = -\$0.25.$$

When the expected value is zero, the game is said to be fair. That is, there is a fifty–fifty chance of winning. When the expected value of a game is negative, it is in favor of the house (i.e., the person or organization running the game). When the expected value of a game is positive, it is in favor of the player. The last situation rarely ever happens unless the con man is not knowledgeable of probability theory.

EXAMPLE: One thousand tickets are sold for \$2 each and there are four prizes. They are \$500, \$250, \$100, and \$50. Find the expected value if a person purchases 2 tickets.

SOLUTION:

Find the expected value if a person purchases one ticket.

Gain, X	**\$499**	**\$249**	**\$99**	**\$49**	**−\$1**
Probability $P(X)$	$\frac{1}{1000}$	$\frac{1}{1000}$	$\frac{1}{1000}$	$\frac{1}{1000}$	$\frac{996}{1000}$

$$E(X) = \$499 \cdot \frac{1}{1000} + \$249 \cdot \frac{1}{1000} + \$99 \cdot \frac{1}{1000} + \$49 \cdot \frac{1}{1000} - \$1 \cdot \frac{996}{1000}$$
$$= -\$0.10$$

The expected value is $-\$0.10$ for one ticket. It is $2(-\$0.10) = -\$0.20$ for two tickets.

Alternate solution

$$E(X) = \$500 \cdot \frac{1}{1000} + \$250 \cdot \frac{1}{1000} + \$100 \cdot \frac{1}{1000} + \$50 \cdot \frac{1}{1000} - \$1$$
$$= -\$0.10$$

$$2(-\$0.10) = -\$0.20$$

Expectation can be used to determine the average amount of money the house can make on each play of a gambling game. Consider the game called Chuck-a-luck. A player pays \$1 and chooses a number from 1 to 6. Then three dice are tossed (usually in a cage). If the player's number comes up once, the player gets \$2. If it comes up twice, the player gets \$3, and if it comes up on all three dice, the player wins \$4. Con men like to say that the probability of any number coming up is $\frac{1}{6}$ on each die; therefore, each number has a probability of $\frac{3}{6}$ or $\frac{1}{2}$ of occurring, and if it occurs more than once, the player wins more money. Hence, the game is in favor of the player. This is not true. The next example shows how to compute the expected value for the game of Chuck-a-luck.

EXAMPLE: Find the expected value for the game Chuck-a-luck.

SOLUTION:

There are $6 \times 6 \times 6 = 216$ outcomes in the sample space for three dice. The probability of winning on each die is $\frac{1}{6}$ and the probability of losing is $\frac{5}{6}$.

The probability that you win on all three dice is $\frac{1}{6} \cdot \frac{1}{6} \cdot \frac{1}{6} = \frac{1}{216}$.

The probability that you lose on all three dice is $\frac{5}{6} \cdot \frac{5}{6} \cdot \frac{5}{6} = \frac{125}{216}$.

The probability that you win on two dice is $\frac{1}{6} \cdot \frac{1}{6} \cdot \frac{5}{6} = \frac{5}{216}$, but this can occur in three different ways: (i) win on the first and the second dice, and lose on the third die, (ii) win on the first die, lose on the second die, and win on the third die, (iii) lose on the first die, and win on the second and third dice. Therefore, the probability of winning on two out of three dice is $3 \cdot \frac{5}{216} = \frac{15}{216}$.

The probability of winning on one die is $\frac{1}{6} \cdot \frac{5}{6} \cdot \frac{5}{6} = \frac{25}{216}$, and there are three different ways to win. Hence, the probability of winning on one die is $3 \cdot \frac{25}{216} = \frac{75}{216}$.

Now the expected value of the game is

X	\$3	\$2	\$1	−\$1
$P(X)$	$\frac{1}{216}$	$\frac{15}{216}$	$\frac{75}{216}$	$\frac{125}{216}$

$$E(X) = \$3 \cdot \frac{1}{216} + \$2 \cdot \frac{15}{216} + \$1 \cdot \frac{75}{216} - \$1 \cdot \frac{125}{216}$$
$$= -\frac{17}{216} \text{ or } -0.078 \text{ or about } -8 \text{ cents.}$$

Hence, on average, the house wins 8 cents on every game played by one player. If 5 people are playing, the house wins about 5×8 cents $= 40$ cents per game.

PRACTICE

1. A box contains five \$1 bills, two \$5 bills, and one \$10 bill. If a person selects one bill at random, find the expected value of the draw.
2. If a person rolls two dice and obtain a sum of 2 or 12, he wins \$15. Find the expectation of the game if the person pays \$5 to play.
3. Five hundred tickets are sold at \$1 each for a color television set worth \$300. Find the expected value if a person purchased one ticket.
4. If a person tosses two coins and gets two heads, the person wins \$10. How much should the person pay if the game is to be fair?
5. A lottery offers one \$1000 prize, two \$500 prizes and five \$100 prizes. Find the expected value of the drawing if 1000 tickets are sold for \$5.00 each and a person purchases one ticket.

ANSWERS

1.

Gain (X)	**\$10**	**\$5**	**\$1**
Probability $P(X)$	$\frac{1}{8}$	$\frac{2}{8}$	$\frac{5}{8}$

$$E(X) = \$10 \cdot \frac{1}{8} + \$5 \cdot \frac{2}{8} + \$1 \cdot \frac{5}{8} = \$3\frac{1}{8} \text{ or } \$3.125$$

2. There is one way to roll a sum of two and one way to roll a sum of 12; there are 36 outcomes in the sample space.

Gain (X)	\$10	−\$5
Probability $P(X)$	$\frac{2}{36}$	$\frac{34}{36}$

$$E(X) = \$10 \cdot \frac{2}{36} + (-\$5) \cdot \frac{34}{36} = -4.17(\text{rounded})$$

Alternate Solution

$$E(X) = \$15 \cdot \frac{2}{36} - \$5 = -\$4.17(\text{rounded})$$

3.

Gain, (X)	\$299	−\$1
Probability $P(X)$	$\frac{1}{500}$	$\frac{499}{500}$

$$E(X) = \$299 \cdot \frac{1}{500} + (-\$1)\frac{499}{500} = -\$0.40$$

Alternate Solution

$$E(X) = \$300 \cdot \frac{1}{500} - \$1 = -\$0.40$$

4. $P(\text{HH}) = \frac{1}{4}$

$$E(X) = 0 = \$10 \cdot \frac{1}{4} - x$$

$$x = \$2.50$$

A person should pay \$2.50 to play the game.

5.

Gain X	\$995	\$495	\$95	−\$5
Probability $P(X)$	$\frac{1}{1000}$	$\frac{2}{1000}$	$\frac{5}{1000}$	$\frac{992}{1000}$

$$E(X) = (\$995)\frac{1}{1000} + \$495\left(\frac{2}{1000}\right) + \$95\left(\frac{5}{1000}\right) + (-\$5)\frac{992}{1000} = -\$2.50$$

Summary

Odds are used to determine the payoffs in gambling games such as lotteries, horse races, and sports betting. Odds are computed from probabilities; however, probabilities can be computed from odds if the true odds are known.

Mathematical expectation can be thought of more or less as a long run average. If the game is played many times, the average of the outcomes or the payouts can be computed using mathematical expectation.

CHAPTER QUIZ

1. Three coins are tossed. What are the odds in favor of getting 3 heads?

   a. 1:7
   b. 3:8
   c. 7:1
   d. 8:3

2. When two dice are rolled, what are the odds against getting doubles?

   a. 1:5
   b. 1:6
   c. 5:1
   d. 6:1

3. When a card is selected from a deck of 52 cards, what are the odds in favor of getting a face card?

   a. 5:2
   b. 3:10
   c. 12:1
   d. 2:5

4. When a die is rolled, what are the odds in favor of getting a 5 or a 6?

   a. 2:3
   b. 1:2
   c. 3:2
   d. 2:1

5. On a roulette wheel, there are 38 numbers: 18 numbers are red, 18 numbers are black, and 2 are green. What are the odds in favor of getting a red number when the ball is rolled?

   a. 19:9
   b. 9:10
   c. 10:9
   d. 9:19

6. If the odds in favor of an event occurring are 3:5, what are the odds against the event occurring?

   a. 5:3
   b. 2:5
   c. 5:8
   d. 8:5

7. If the odds against an event occurring are 8:3, what are the odds in favor of the event occurring?

   a. 3:8
   b. 11:3
   c. 11:8
   d. 8:5

8. The probability of an event occurring is $\frac{5}{13}$. What are the odds in favor of the event occurring?

   a. 8:5
   b. 13:5
   c. 5:13
   d. 5:8

9. The probability of an event occurring is $\frac{7}{12}$. What are the odds against the event occurring?

   a. 5:12
   b. 12:5
   c. 5:7
   d. 7:5

10. What are the odds for a fair game?

    a. 0:0
    b. 1:1
    c. 2:1
    d. 1:2

11. When a game is fair the expected value would be

    a. 1
    b. 0
    c. −1
    d. 0.5

12. When three coins are tossed, what is the expected value of the number of heads?

    a. 1
    b. 2
    c. 1.5
    d. 2.5

13. A special die is made with 1 one, 2 twos, 3 threes. What is the expected number of spots for one roll?

    a. $3\frac{1}{2}$
    b. $1\frac{2}{3}$
    c. 2
    d. $2\frac{1}{3}$

14. Two hundred raffle tickets are sold for $1.00 each. One prize of $75 is awarded. What is the expected value if a person purchases one ticket?

    a. $0.625
    b. −$0.625
    c. $1.75
    d. −$1.75

15. A box contains 5 one dollar bills, 3 five dollar bills, and 2 ten dollar bills. A person selects one bill at random and wins that bill. How much should the person pay to play the game if it is to be fair?

    a. $2.00
    b. $5.00
    c. $3.00
    d. $4.00

Probability Sidelight

PROBABILITY AND GENETICS

An Austrian botanist, Gregor Mendel (1822–1884), studied genetics and used probability theory to verify his results. Mendel lived in a monastery all of his adult life and based his research on the observation of plants. He published his results in an obscure journal and the results remained unknown until the beginning of the 20th century. At that time, his research was used by a mathematician, G. H. Hardy, to study human genetics.

Genetics is somewhat more complicated than what is presented here. However, what is important here is to explain how probability is used in genetics.

One of Mendel's studies was on the color of the seeds of pea plants. There were two colors, yellow and green. Mendel theorized that each egg cell and each pollen cell contained two color genes that split on fertilization. The offspring then contained one gene cell from each donor. There were three possibilities: pure yellow seeds, pure green seeds, and hybrid-yellow seeds. The pure yellow seeds contain two yellow genes. The pure green seeds contained two green genes. The hybrid-yellow seeds contain one yellow gene and one green gene. This seed was yellow since the yellow gene is dominant over the green gene. The green gene is said to be recessive.

Next consider the possibilities. If there are two pure yellow plants, then the results of fertilization will be YY as shown.

	Y	Y
Y	YY	YY
Y	YY	YY

Hence, $P(\mathrm{YY}) = 1$.

The results of two pure green plants will be gg.

	g	g
g	gg	gg
g	gg	gg

Hence, $P(\mathrm{gg}) = 1$.

What happens with a pure yellow and a pure green plant?

	g	g
Y	Yg	Yg
Y	Yg	Yg

Hence, $P(\mathrm{Yg}) = 1$.

What happens with two hybrid yellow plants?

	Y	g
Y	YY	Yg
g	gY	gg

Hence, $P(\text{YY}) = \frac{1}{4}$, $P(\text{Yg}) = P(\text{gY}) = \frac{1}{2}$, and $\text{P(gg)} = \frac{1}{4}$.

What about a pure yellow plant and a hybrid yellow plant?

	Y	g
Y	YY	Yg
Y	YY	Yg

Hence, $P(\text{YY}) = \frac{1}{2}$ and $P(\text{Yg}) = \frac{1}{2}$.

What about a hybrid yellow plant and a pure green plant?

	g	g
Y	Yg	Yg
g	gg	gg

Hence $P(\text{Yg}) = \frac{1}{2}$ and $P(\text{gg}) = \frac{1}{2}$.

This format can be used for other traits such as gender, eye color, etc. For example, for the gender of children, the female egg contains two X chromosomes, and the male contains X and Y chromosomes. Hence the results of fertilization are

	X	Y
X	XX	XY
X	XX	XY

Hence, $P(\text{female}) = P(\text{XX}) = \frac{1}{2}$ and $P(\text{male}) = P(\text{XY}) = \frac{1}{2}$.

Mendel performed experiments and compared the results with the theoretical probability of the outcomes in order to verify his hypotheses.

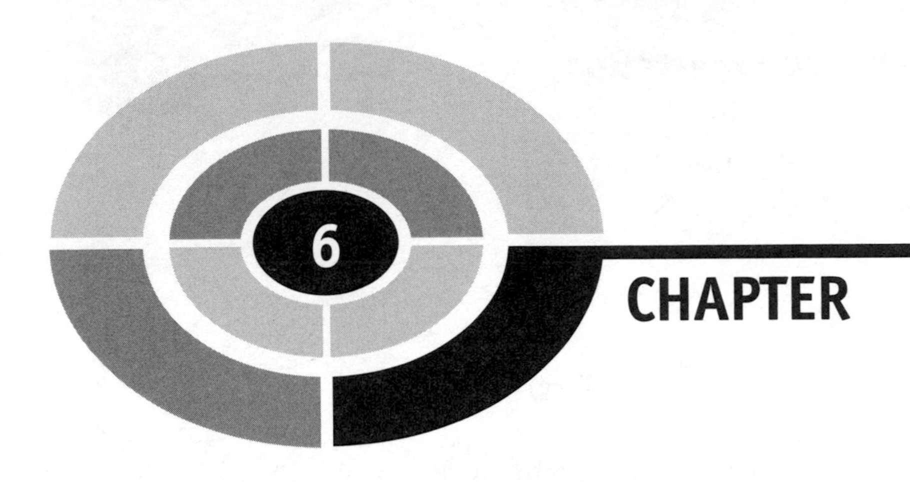

The Counting Rules

Introduction

Since probability problems require knowing the total number of ways one or more events can occur, it is necessary to have a way to compute the number of outcomes in the sample spaces for a probability experiment. This is especially true when the number of outcomes is large. For example, when finding the probability of a specific poker hand, it is necessary to know the number of different possible ways five cards can be dealt from a 52-card deck. (This computation will be shown later in this chapter.)

In order to do the computation, we use the fundamental counting rule, the permutation rules, and the combination rule. The rules then can be used to compute the probability for events such as winning lotteries, getting a specific hand in poker, etc.

The Fundamental Counting Rule

The first rule is called the **Fundamental Counting Rule**.

For a sequence of n events in which the first event can occur in k_1 ways and the second event can occur in k_2 ways and the third event can occur in k_3 ways, and so on, the total number of ways the sequence can occur is $k_1 \cdot k_2 \cdot k_3 \ldots k_n$.

EXAMPLE: In order to paint a room, a person has a choice of four colors: white, light blue, yellow, and light green; two types of paint: oil or latex; and three types of texture: flat, semi-glass, or satin. How many different selections can be made?

SOLUTION:

There are four colors, two types of paint, and three textures, so the total number of ways a paint can be selected is $4 \cdot 2 \cdot 3 = 24$ ways.

EXAMPLE: There are four blood types A, B, AB, and O. Blood can be Rh^+ or Rh^-. Finally, a donor can be male or female. How many different classifications can be made?

SOLUTION:

$$4 \cdot 2 \cdot 2 = 16$$

When determining the number of different ways a sequence of events can occur, it is necessary to know whether or not repetitions are permitted. The next two examples show the difference between the two situations.

EXAMPLE: The employees of a company are given a 4-digit identification number. How many different numbers are available if repetitions are permitted?

SOLUTION:

There are 10 digits (zero through nine), so each of the four digits can be selected in ten different ways since repetitions are permitted. Hence the total number of identification numbers is $10 \cdot 10 \cdot 10 \cdot 10 = 10^4 = 10{,}000$.

EXAMPLE: The employees of a company are given 4-digit identification numbers; however, repetitions are not allowed. How many different numbers are available?

SOLUTION:

In this case, there are 10 ways to select the first digit, 9 ways to select the second digit, 8 ways to select the third digit, and 7 ways to select the fourth digit, so the total number of ways is $10 \cdot 9 \cdot 8 \cdot 7 = 5040$.

PRACTICE

1. A person can select eight different colors for an automobile body, five different colors for the interior, and white or black sidewall tires. How many different color combinations are there for the automobile?
2. A person can select one of five different colors for brick borders, one type of six different ground coverings, and one of three different types of shrubbery. How many different types of landscape designs are there?
3. How many different types of identification cards consisting of 4 letters can be made from the first five letters of the alphabet if repetitions are allowed?
4. How many different types of identification cards consisting of 4 letters can be made from the first 5 letters of the alphabet if repetitions are not allowed?
5. A license plate consists of 2 letters and 3 digits. How many different plates can be made if repetitions are permitted? How many can be made if repetitions are not permitted?

ANSWERS

1. $8 \cdot 5 \cdot 2 = 80$ color combinations
2. $5 \cdot 6 \cdot 3 = 90$ types
3. $5 \cdot 5 \cdot 5 \cdot 5 = 5^4 = 625$ cards
4. $5 \cdot 4 \cdot 3 \cdot 2 = 120$ cards
5. Repetitions permitted: $26 \cdot 26 \cdot 10 \cdot 10 \cdot 10 = 676{,}000$

   Repetitions not permitted $26 \cdot 25 \cdot 10 \cdot 9 \cdot 8 = 468{,}000$

Factorials

In mathematics there is a notation called **factorial notation**, which uses the exclamation point. Some examples of factorial notation are

$$6! = 6 \cdot 5 \cdot 4 \cdot 3 \cdot 2 \cdot 1 = 720$$
$$3! = 3 \cdot 2 \cdot 1 = 6$$
$$5! = 5 \cdot 4 \cdot 3 \cdot 2 \cdot 1 = 120$$
$$1! = 1$$

Notice that factorial notation means to start with the number and find its product with all of the whole numbers less than the number and stopping at one. Formally defined,

$$n! = n \cdot (n-1) \cdot (n-2) \ldots 3 \cdot 2 \cdot 1$$

Factorial notation can be stopped at any time. For example,

$$6! = 6 \cdot 5! = 6 \cdot 5 \cdot 4!$$
$$10! = 10 \cdot 9! = 10 \cdot 9 \cdot 8!$$

In order to use the formulas in the rest of the chapter, it is necessary to know how to multiply and divide factorials. In order to multiply factorials, it is necessary to multiply them out and then multiply the products. For example,

$$3! \cdot 4! = 3 \cdot 2 \cdot 1 \cdot 4 \cdot 3 \cdot 2 \cdot 1 = 144$$

Notice $3! \cdot 4! \neq 12!$ Since $12! = 479{,}001{,}600$

EXAMPLE: Find the product of $5! \cdot 4!$

SOLUTION:

$$5! \cdot 4! = 5 \cdot 4 \cdot 3 \cdot 2 \cdot 1 \cdot 4 \cdot 3 \cdot 2 \cdot 1 = 2880$$

Division of factorials is somewhat tricky. You can always multiply them out and then divide the top number by the bottom number. For example,

$$\frac{8!}{6!} = \frac{8 \cdot 7 \cdot 6 \cdot 5 \cdot 4 \cdot 3 \cdot 2 \cdot 1}{6 \cdot 5 \cdot 4 \cdot 3 \cdot 2 \cdot 1} = \frac{40{,}320}{720} = 56$$

or

you can cancel out, as shown:

$$\frac{8!}{6!} = \frac{8 \cdot 7 \cdot \cancel{6!}}{\cancel{6!}} = 8 \cdot 7 = 56$$

You cannot divide factorials directly:

$$\frac{8!}{4!} \neq 2! \text{ since } 8! = 40{,}320 \text{ and } 4! = 24, \text{ then } \frac{40{,}320}{24} = 1680$$

EXAMPLE: Find the quotient $\frac{7!}{3!}$

SOLUTION:

$$\frac{7!}{3!} = \frac{7 \cdot 6 \cdot 5 \cdot 4 \cdot \cancel{3!}}{\cancel{3!}} = 7 \cdot 6 \cdot 5 \cdot 4 = 840$$

Most scientific calculators have a factorial key. It is the key with "!". Also $0! = 1$ by definition.

PRACTICE

Find the value of each

1. $2!$
2. $7!$
3. $9!$
4. $4!$
5. $6! \cdot 3!$
6. $4! \cdot 8!$
7. $7! \cdot 2!$
8. $\frac{10!}{8!}$
9. $\frac{5!}{2!}$
10. $\frac{6!}{3!}$

SOLUTIONS

1. $2! = 2 \cdot 1 = 2$
2. $7! = 7 \cdot 6 \cdot 5 \cdot 4 \cdot 3 \cdot 2 \cdot 1 = 5040$
3. $9! = 9 \cdot 8 \cdot 7 \cdot 6 \cdot 5 \cdot 4 \cdot 3 \cdot 2 \cdot 1 = 362{,}880$
4. $4! = 4 \cdot 3 \cdot 2 \cdot 1 = 24$
5. $6! \cdot 3! = 6 \cdot 5 \cdot 4 \cdot 3 \cdot 2 \cdot 1 \cdot 3 \cdot 2 \cdot 1 = 4320$
6. $4! \cdot 8! = 4 \cdot 3 \cdot 2 \cdot 1 \cdot 8 \cdot 7 \cdot 6 \cdot 5 \cdot 4 \cdot 3 \cdot 2 \cdot 1 = 967{,}680$
7. $7! \cdot 2! = 7 \cdot 6 \cdot 5 \cdot 4 \cdot 3 \cdot 2 \cdot 1 \cdot 2 \cdot 1 = 10{,}080$
8. $\dfrac{10!}{8!} = \dfrac{10 \cdot 9 \cdot \cancel{8!}}{\cancel{8!}} = 10 \cdot 9 = 90$
9. $\dfrac{5!}{2!} = \dfrac{5 \cdot 4 \cdot 3 \cdot \cancel{2!}}{\cancel{2!}} = 5 \cdot 4 \cdot 3 = 60$
10. $\dfrac{6!}{3!} = \dfrac{6 \cdot 5 \cdot 4 \cdot \cancel{3!}}{\cancel{3!}} = 6 \cdot 5 \cdot 4 = 120$

The Permutation Rules

The second way to determine the number of outcomes of an event is to use the **permutation rules.** An arrangement of n distinct objects in a specific order is called a **permutation**. For example, if an art dealer had 3 paintings, say A, B, and C, to arrange in a row on a wall, there would be 6 distinct ways to display the paintings. They are

ABC BAC CAB
ACB BCA CBA

The total number of different ways can be found using the fundamental counting rule. There are 3 ways to select the first object, 2 ways to select the second object, and 1 way to select the third object. Hence, there are $3 \cdot 2 \cdot 1 = 6$ different ways to arrange three objects in a row on a shelf.

Another way to solve this kind of problem is to use permutations. The number of permutations of n objects using all the objects is $n!$.

EXAMPLE: In how many different ways can 6 people be arranged in a row for a photograph?

SOLUTION:

This is a permutation of 6 objects. Hence $6! = 6 \cdot 5 \cdot 4 \cdot 3 \cdot 2 \cdot 1 = 720$ ways.

In the previous example, all the objects were used; however, in many situations only some of the objects are used. In this case, the **permutation rule** can be used.

The arrangement of n objects in a specific order using r objects at a time is called a **permutation** of n objects taking r objects at a time. It is written as $_nP_r$ and the formula is

$$_nP_r = \frac{n!}{(n-r)!}$$

EXAMPLE: In how many different ways can 3 people be arranged in a row for a photograph if they are selected from a group of 5 people?

SOLUTION:

Since 3 people are being selected from 5 people and arranged in a specific order, $n = 5$, $r = 3$. Hence, there are

$$_5P_3 = \frac{5!}{(5-3)!} = \frac{5!}{2!} = \frac{5 \cdot 4 \cdot 3 \cdot \cancel{2!}}{\cancel{2!}} = 5 \cdot 4 \cdot 3 = 60 \text{ ways}$$

EXAMPLE: In how many different ways can a chairperson and secretary be selected from a committee of 9 people?

SOLUTION:

In this case, $n = 9$ and $r = 2$. Hence, there are $_9P_2$ ways of selecting two people to fill the two positions.

$$_9P_2 = \frac{9!}{(9-2)!} = \frac{9!}{7!} = \frac{9 \cdot 8 \cdot \cancel{7!}}{\cancel{7!}} = 72 \text{ ways}$$

EXAMPLE: How many different signals can be made from seven different flags if four flags are displayed in a row?

SOLUTION:

Hence $n=7$ and $r=4$, so

$$_7P_4 = \frac{7!}{(7-4)!} = \frac{7!}{3!} = \frac{7 \cdot 6 \cdot 5 \cdot 4 \cdot \cancel{3!}}{\cancel{3!}} = 7 \cdot 6 \cdot 5 \cdot 4 = 840$$

In the preceding examples, all the objects were different, but when some of the objects are identical, the second permutation rule can be used.

The number of permutations of n objects when r_1 objects are identical, r_2 objects are identical, etc. is

$$\frac{n!}{r_1! r_2! \ldots r_p!}$$

where $r_1 + r_2 + \ldots + r_p = n$

EXAMPLE: How many different permutations can be made from the letters of the word **Mississippi**?

SOLUTION:

There are 4 s, 4 i, 2 p, and 1 m; hence, $n=11$, $r_1=4$, $r_2=4$, $r_3=2$, and $r_4=1$

$$\frac{11!}{4! \cdot 4! \cdot 2! \cdot 1!} = \frac{11 \cdot 10 \cdot 9 \cdot 8 \cdot 7 \cdot 6 \cdot 5 \cdot \cancel{4!}}{\cancel{4!} \cdot 4 \cdot 3 \cdot 2 \cdot 1 \cdot 2 \cdot 1 \cdot 1} = \frac{1{,}663{,}200}{48} = 34{,}650$$

EXAMPLE: An automobile dealer has 3 Fords, 2 Buicks, and 4 Dodges to place in the front row of his car lot. In how many different ways by make of car can he display the automobiles?

SOLUTION:

Let $n=3+2+4=9$ automobiles; $r_1=3$ Fords, $r_2=2$ Buicks, and $r_3=4$ Dodges; then there are $\frac{9!}{3! \cdot 2! \cdot 4!} = \frac{9 \cdot 8 \cdot 7 \cdot 6 \cdot 5 \cdot \cancel{4!}}{3 \cdot 2 \cdot 1 \cdot 2 \cdot 1 \cdot \cancel{4!}} = 1260$ ways to display the automobiles.

PRACTICE

1. How many different batting orders can a manager make with his starting team of 9 players?
2. In how many ways can a nurse select three patients from 8 patients to visit in the next hour? The order of visitation is important.
3. In how many different ways can a president, vice-president, secretary, and a treasurer be selected from a club with 15 members?
4. In how many different ways can an automobile repair shop owner select five automobiles to be repaired if there are 8 automobiles needing service? The order is important.
5. How many different signals using 6 flags can be made if 3 are red, 2 are blue, and 1 is white?

ANSWERS

1. $9! = 9 \cdot 8 \cdot 7 \cdot 6 \cdot 5 \cdot 4 \cdot 3 \cdot 2 \cdot 1 = 362{,}880$
2. $_8P_3 = \frac{8!}{(8-3)!} = \frac{8!}{5!} = \frac{8 \cdot 7 \cdot 6 \cdot \cancel{5!}}{\cancel{5!}} = 336$
3. $_{15}P_4 = \frac{15!}{(15-4)!} = \frac{15!}{11!} = \frac{15 \cdot 14 \cdot 13 \cdot 12 \cdot \cancel{11!}}{\cancel{11!}} = 32{,}760$
4. $_8P_5 = \frac{8!}{(8-5)!} = \frac{8!}{3!} = \frac{8 \cdot 7 \cdot 6 \cdot 5 \cdot 4 \cdot \cancel{3!}}{\cancel{3!}} = 6720$
5. $\frac{6!}{3!2!1!} = \frac{6 \cdot 5 \cdot 4 \cdot \cancel{3!}}{\cancel{3!} \cdot 2 \cdot 1 \cdot 1} = 60$

Combinations

Sometimes when selecting objects, the order in which the objects are selected is not important. For example, when five cards are dealt in a poker game, the order in which you receive the cards is not important. When 5 balls are selected in a lottery, the order in which they are selected is not important. These situations differ from permutations in which order is important and are called combinations. A **combination** is a selection of objects without regard to the order in which they are selected.

Suppose two letters are selected from the four letters, A, B, C, and D. The different permutations are shown on the left and the different combinations are shown on the right.

PERMUTATIONS				COMBINATIONS	
AB	BA	CA	DA	AB	BC
AC	BC	CB	DB	AC	BD
AD	BD	CD	DC	AD	CD

Notice that in a permutation AB differs from BA, but in a combination AB is the same as BA. The **combination rule** is used to find the number of ways to select objects without regard to order.

The number of ways of selecting r objects from n objects without regard to order is

$$_{n}C_{r} = \frac{n!}{(n-r)!r!}$$

Note: The symbol $_{n}C_{r}$ is used for combinations; however, some books use other symbols. Two of the most commonly used symbols are C_{r}^{n} or $\binom{n}{r}$.

EXAMPLE: In how many ways can 2 objects be selected from 6 objects without regard to order?

SOLUTION:

Let $n = 6$ and $r = 2$,

$$_{6}C_{2} = \frac{6!}{(6-2)!2!} = \frac{6!}{4!2!} = \frac{6 \cdot 5 \cdot \cancel{4!}}{\cancel{4!} \cdot 2 \cdot 1} = 15$$

EXAMPLE: A salesperson has to visit 10 stores in a large city. She decides to visit 6 stores on the first day. In how many different ways can she select the 6 stores? The order is not important.

SOLUTION:

Let $n = 10$ and $r = 6$; then

$$_{10}C_{6} = \frac{10!}{(10-6)!6!} = \frac{10!}{4!6!} = \frac{10 \cdot 9 \cdot 8 \cdot 7 \cdot \cancel{6!}}{4 \cdot 3 \cdot 2 \cdot 1 \cdot \cancel{6!}} = 210$$

She can select the 6 stores in 210 ways.

EXAMPLE: In a classroom, there are 8 women and 5 men. A committee of 3 women and 2 men is to be formed for a project. How many different possibilities are there?

SOLUTION:

In this case, you must select 3 women from 8 women and 2 men from 5 men. Since the word "and" is used, multiply the answers.

$$
\begin{aligned}
{}_8C_3 \cdot {}_5C_2 &= \frac{8!}{(8-3)!3!} \cdot \frac{5!}{(5-2)!2!} \\
&= \frac{8!}{5! \cdot 3!} \cdot \frac{5!}{3! \cdot 2!} \\
&= \frac{8 \cdot 7 \cdot 6 \cdot \cancel{5!}}{\cancel{5!} \cdot 3 \cdot 2 \cdot 1} \cdot \frac{5 \cdot 4 \cdot \cancel{3!}}{\cancel{3!} \cdot 2 \cdot 1} = 56 \cdot 10 \\
&= 560
\end{aligned}
$$

Hence, there are 560 different ways to make the selection.

PRACTICE

1. In how many ways can a large retail store select 3 sites on which to build a new store if it has 12 sites to choose from?
2. In how many ways can Mary select two friends to go to a movie with if she has 7 friends to choose from?
3. In how many ways can a real estate agent select 10 properties to place in an advertisement if she has 15 listings to choose from?
4. In how many ways can a committee of 3 elementary school teachers be selected from a school district which has 8 elementary school teachers?
5. In a box of 10 calculators, one is defective. In how many ways can four calculators be selected if the defective calculator is included in the group?

ANSWERS

1. $n=12, r=3$

$$_{12}C_3 = \frac{12!}{(12-3)!3!} = \frac{12!}{9!3!} = \frac{12 \cdot 11 \cdot 10 \cdot \cancel{9!}}{\cancel{9!} \cdot 3 \cdot 2 \cdot 1} = 220$$

2. $n=7, r=2$

$$_7C_2 = \frac{7!}{(7-2)!2!} = \frac{7!}{5!2!} = \frac{7 \cdot 6 \cdot \cancel{5!}}{\cancel{5!} \cdot 2 \cdot 1} = 21$$

3. $n=15, r=10$

$$_{15}C_{10} = \frac{15!}{(15-10)!10!} = \frac{15!}{5!10!} = \frac{15 \cdot 14 \cdot 13 \cdot 12 \cdot 11 \cdot \cancel{10!}}{5 \cdot 4 \cdot 3 \cdot 2 \cdot 1 \cdot \cancel{10!}} = 3003$$

4. $n=8, r=3$

$$_8C_3 = \frac{8!}{(8-3)!3!} = \frac{8!}{5!3!} = \frac{8 \cdot 7 \cdot 6 \cdot \cancel{5!}}{\cancel{5!} \cdot 3 \cdot 2 \cdot 1} = 56 \text{ ways}$$

5. If the defective calculator is included, then you must select the other calculators from the remaining 9 calculators; hence, there are $_9C_3$ ways to select the 4 calculators including the defective calculator.

$$_9C_3 = \frac{9!}{(9-3)!3!} = \frac{9!}{6!3!} = \frac{9 \cdot 8 \cdot 7 \cdot \cancel{6!}}{\cancel{6!} \cdot 3 \cdot 2 \cdot 1} = 84 \text{ ways}$$

Probability and the Counting Rules

A wide variety of probability problems can be solved using the counting rules and the probability rule.

EXAMPLE: Find the probability of getting a flush (including a straight flush) when 5 cards are dealt from a deck of 52 cards.

SOLUTION:

A flush consists of 5 cards of the same suit. That is, either 5 clubs or 5 spades or 5 hearts or 5 diamonds, and includes straight flushes.

Since there are 13 cards in a suit, there are ${}_{13}C_5$ ways to get a flush in one suit, and there are 4 suits, so the number of ways to get a flush is

$$\begin{aligned} 4 \cdot {}_{13}C_5 &= 4 \cdot \frac{13!}{(13-5)!5!} = 4 \cdot \frac{13!}{8!5!} \\ &= 4 \cdot \frac{13 \cdot 12 \cdot 11 \cdot 10 \cdot 9 \cdot \cancel{8!}}{\cancel{8!} \cdot 5 \cdot 4 \cdot 3 \cdot 2 \cdot 1} \\ &= 5148 \end{aligned}$$

There are ${}_{52}C_5$ ways to select 5 cards.

$${}_{52}C_5 = \frac{52!}{(52-5)!5!} = \frac{52!}{47!5!} = \frac{52 \cdot 51 \cdot 50 \cdot 49 \cdot 48 \cdot \cancel{47!}}{\cancel{47!} \cdot 5 \cdot 4 \cdot 3 \cdot 2 \cdot 1} = 2{,}598{,}960$$

$P(\text{flush}) = \dfrac{5148}{2{,}598{,}960} = 0.00198$ or about 0.002, which is about one chance in 500.

EXAMPLE: A student has a choice of selecting three elective courses for the next semester. He can choose from six humanities or four psychology courses. Find the probability that all three courses selected will be humanities courses assuming he selects them at random.

SOLUTION:

Since there are six humanities courses, and the student needs to select three of them, there are ${}_6C_3$ ways of doing this:

$${}_6C_3 = \frac{6!}{(6-3)!3!} = \frac{6!}{3!3!} = \frac{6 \cdot 5 \cdot 4 \cdot \cancel{3!}}{\cancel{3!} \cdot 3 \cdot 2 \cdot 1} = 20$$

The total number of ways of selecting 3 courses from 10 courses is ${}_{10}C_3$.

$${}_{10}C_3 = \frac{10!}{(10-3)!3!} = \frac{10!}{7!3!} = \frac{10 \cdot 9 \cdot 8 \cdot \cancel{7!}}{\cancel{7!} \cdot 3 \cdot 2 \cdot 1} = 120$$

Hence, the probability of selecting all humanities courses is

$$\frac{20}{120} = \frac{1}{6} \approx 0.167$$

There is one chance in 6 that he will select all humanities courses if he chooses them at random.

EXAMPLE: An identification card consists of 3 digits selected from 10 digits. Find the probability that a randomly selected card contains the digits 1, 2, and 3. Repetitions are not permitted.

SOLUTION:

The number of permutations of 1, 2, and 3 is ${}_3P_3 = \frac{3!}{(3-3)!} = \frac{3!}{0!} = \frac{3 \cdot 2 \cdot 1}{1} = 6$

The number of permutations of 3 digits each that can be made from 10 digits is

$${}_{10}P_3 = \frac{10!}{(10-3)!} = \frac{10!}{7!} = \frac{10 \cdot 9 \cdot 8 \cdot \cancel{7!}}{\cancel{7!}} = 720$$

Hence the probability that the card contains 1, 2, and 3 in any order is

$$\frac{6}{720} = \frac{1}{120} \approx 0.008$$

PRACTICE

1. In a classroom, there are 10 men and 6 women. If 3 students are selected at random to give a presentation, find the probability that all 3 are women.
2. A carton contains 12 toasters, 3 of which are defective. If four toasters are sold at random, find the probability that exactly one will be defective.
3. If 100 tickets are sold for two prizes, and one person buys two tickets, find the probability that that person wins both prizes.
4. A committee of 3 people is formed from 6 nurses and 4 doctors. Find the probability that the committee contains 2 nurses and one doctor. The committee members are selected at random.
5. If 5 cards are dealt, find the probability of getting 4 of a kind.

ANSWERS

1. There are ${}_6C_3$ ways to select 3 women from 6 women.

$${}_6C_3 = \frac{6!}{(6-3)!3!} = \frac{6!}{3!3!} = \frac{6 \cdot 5 \cdot 4 \cdot \cancel{3!}}{\cancel{3!} \cdot 3 \cdot 2 \cdot 1} = 20$$

There are ${}_{16}C_3$ ways to select 3 people from 16 people.

$${}_{16}C_3 = \frac{16!}{(16-3)!3!} = \frac{16!}{13!3!} = \frac{16 \cdot 15 \cdot 14 \cdot \cancel{13!}}{\cancel{13!} \cdot 3 \cdot 2 \cdot 1} = 560$$

$$P(\text{3 women}) = \frac{20}{560} = \frac{1}{28} \approx 0.036$$

2. There are ${}_3C_1$ ways to select a defective toaster and ${}_9C_3$ ways to select 3 nondefective toasters.

$${}_3C_1 \cdot {}_9C_3 = \frac{3!}{(3-1)!1!} \cdot \frac{9!}{(9-3)!3!} = \frac{3!}{2!1!} \cdot \frac{9!}{6!3!} = 3 \cdot 84 = 252$$

There are ${}_{12}C_4$ ways to select 4 toasters from 12 toasters.

$${}_{12}C_4 = \frac{12!}{(12-4)!4!} = \frac{12!}{8!4!} = \frac{12 \cdot 11 \cdot 10 \cdot 9 \cdot \cancel{8}}{\cancel{8}! \cdot 4 \cdot 3 \cdot 2 \cdot 1} = 495$$

Hence, $P(\text{1 defective toaster and 3 nondefective toasters}) = \frac{252}{495} = \frac{28}{55} \approx 0.509$

3. There are ${}_2C_2$ ways to win 2 prizes.

$${}_2C_2 = \frac{2!}{(2-2)!2!} = \frac{2!}{2!} = 1$$

There are ${}_{100}C_2$ ways to give away the prizes

$${}_{100}C_2 = \frac{100!}{(100-2)!2!} = \frac{100 \cdot 99 \cdot \cancel{98!}}{\cancel{98!} \cdot 2 \cdot 1} = 4950$$

$$P(\text{winning both prizes}) = \frac{1}{4950} = 0.0002$$

4. There are ${}_6C_2$ ways to select the nurses, and ${}_4C_1$ ways to select the doctors.

$${}_6C_2 \cdot {}_4C_1 = \frac{6!}{4!2!} \cdot \frac{4!}{3!1!} = 15 \cdot 4 = 60$$

There are ${}_{10}C_3$ ways to select 3 people.

$${}_{10}C_3 = \frac{9!}{(9-3)!3!} = \frac{9!}{6!3!} = 84 \quad \frac{10!}{(10-3)!3!} = \frac{10!}{7!3!} = 120$$

$$P(\text{2 nurses and 1 doctor}) = \frac{60}{120} = \frac{1}{2}$$

5. There are 13 cards in each suit; hence, there are 13 ways to get 4 of a kind and 48 ways to get the fifth card. Therefore, there are 624 ways to get 4 of a kind. There are ${}_{52}C_5$ ways to deal 5 cards.

$${}_{52}C_5 = \frac{52!}{47!5!} = 2{,}598{,}960$$

Hence,

$$P(\text{4 of a kind}) = \frac{624}{2{,}598{,}960} = \frac{13}{54{,}145} \approx 0.0002$$

Summary

In order to determine the number of outcomes of events, the fundamental counting rule, the permutation rules, and the combination rule can be used. The difference between a permutation and a combination is that for a permutation, the order or arrangement of the objects is important. For example, order is important in phone numbers, identification tags, social security numbers, license plates, etc. Order is not important when selecting objects from a group. Many probability problems can be solved by using the counting rules to determine the number of outcomes of the events that are used in the problems.

CHAPTER QUIZ

1. The value of 6! is
   a. 6
   b. 30
   c. 120
   d. 720

2. The value of 0! is
   a. 0
   b. 1
   c. 10
   d. 100

3. The value of ${}_8P_3$ is
   a. 120
   b. 256
   c. 336
   d. 432

4. The value of ${}_5C_2$ is
   a. 10
   b. 12
   c. 120
   d. 324

5. The number of 3-digit telephone area codes that can be made if repetitions are not allowed is
   a. 100
   b. 720
   c. 1000
   d. 504

6. In how many different ways can a person select one book from 3 novels, one book from 5 biographies and one book from 7 self-help books.
   a. 15
   b. 105
   c. 3
   d. 22

7. In how many ways can 7 different calculators be displayed in a row on a shelf?
   a. 7
   b. 49
   c. 823,543
   d. 5,040

8. If a board of directors consists of 10 people, in how many ways can a chief executive officer, a director, a treasurer, and a secretary be selected?
   a. 5040
   b. 210
   c. 40
   d. 14

9. How many different flag signals can be made from 3 red flags, 2 green flags, and 2 white flags?
   a. 49
   b. 84
   c. 210
   d. 320

10. In how many ways can 3 boxes of cereal be selected for testing from 12 boxes?
    a. 36
    b. 220
    c. 480
    d. 1320

11. In how many ways can a jury of 5 men and 7 women be selected from 10 men and 10 women?
    a. 120
    b. 252
    c. 372
    d. 30,240

12. A phone extension consists of 3 digits. If all the digits have a probability of being selected, what is the probability that the extension consists of the digits 1, 2, and 3 in any order? Repetitions are allowed.
    a. 0.555
    b. 0.006
    c. 0.233
    d. 0.125

13. Three cards are selected at random; what is the probability that they will all be clubs?
    a. 0.002
    b. 0.034
    c. 0.013
    d. 0.127

14. At a used book sale, there are 6 novels and 4 biographies. If a person selects 4 books at random, what is the probability that the person selects two novels and two biographies?
    a. 0.383
    b. 0.562
    c. 0.137
    d. 0.429

15. To win a lottery, a person must select 4 numbers in any order from 20 numbers. Repetitions are not allowed. What is the probability that the person wins?
    a. 0.0002
    b. 0.0034
    c. 0.0018
    d. 0.0015

Probability Sidelight

THE CLASSICAL BIRTHDAY PROBLEM

What do you think the chances are that in a classroom of 23 students, two students would have the same birthday (day and month)? Most people would think the probability is very low since there are 365 days in a year; however, the probability is slightly greater than 50%! Furthermore, as the number of students increases, the probability increases very rapidly. For example, if there are 30 students in the room, there is a 70% chance that two students will have the same birthday, and when there are 50 students in the room, the probability jumps to 97%!

The problem can be solved by using permutations and the probability rules. It must be assumed that all birthdays are equally likely. This is not necessarily true, but it has little effect on the solution. The way to solve the problem is to find the probability that no two people have the same birthday and subtract it from one. Recall $P(E) = 1 - P(\overline{E})$.

For example, suppose that there were only three people in the room. Then the probability that each would have a different birthday would be

$$\left(\frac{365}{365}\right) \cdot \left(\frac{364}{365}\right) \cdot \left(\frac{363}{365}\right) = \frac{{}_{365}P_3}{(365)^3} = 0.992$$

The reasoning here is that the first person could be born on any day of the year. Now if the second person would have a different birthday, there are

364 days left, so the probability that the second person was born on a different day is $\frac{364}{365}$. The reasoning is the same for the next person. Now since the probability is 0.992 that the three people have different birthdays, the probability that any two have the same birthday is $1 - 0.992 = 0.008$ or 0.8%.

In general, in a room with k people, the probability that at least two people will have the same birthday is $1 - \frac{_{365}P_k}{365^k}$.

In a room with 23 students, then, the probability that at least two students will have the same birthday is $1 - \frac{_{365}P_{23}}{365^{23}} = 0.507$ or 50.7%.

It is interesting to note that two presidents, James K. Polk and Warren G. Harding, were both born on November 2. Also, John Adams and Thomas Jefferson both died on July 4. What is even more unusual is that they both died on the same day of the same year, July 4, 1826. Another President, James Monroe, also died on July 4, but the year was 1831.

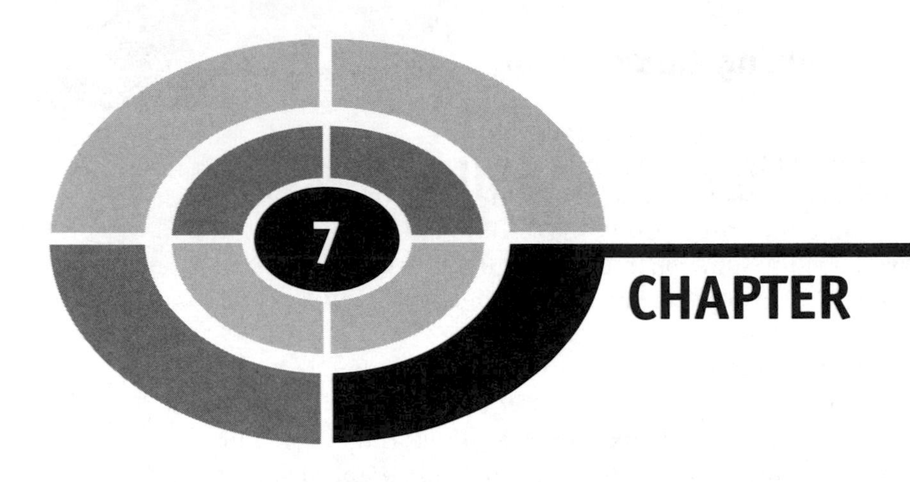

The Binomial Distribution

Introduction

Many probability problems involve assigning probabilities to the outcomes of a probability experiment. These probabilities and the corresponding outcomes make up a *probability distribution*. There are many different probability distributions. One special probability distribution is called the *binomial distribution*. The binomial distribution has many uses such as in gambling, in inspecting parts, and in other areas.

Discrete Probability Distributions

In mathematics, a **variable** can assume different values. For example, if one records the temperature outside every hour for a 24-hour period, temperature is considered a variable since it assumes different values. Variables whose values are due to chance are called **random variables.** When a die is rolled, the value of the spots on the face up occurs by chance; hence, the number of spots on the face up on the die is considered to be a random variable. The outcomes of a die are 1, 2, 3, 4, 5, and 6, and the probability of each outcome occurring is $\frac{1}{6}$. The outcomes and their corresponding probabilities can be written in a table, as shown, and make up what is called a probability distribution.

Value, x	1	2	3	4	5	6
Probability, $P(x)$	$\frac{1}{6}$	$\frac{1}{6}$	$\frac{1}{6}$	$\frac{1}{6}$	$\frac{1}{6}$	$\frac{1}{6}$

A **probability distribution** consists of the values of a random variable and their corresponding probabilities.

There are two kinds of probability distributions. They are *discrete* and *continuous*. A **discrete** variable has a countable number of values (countable means values of zero, one, two, three, etc.). For example, when four coins are tossed, the outcomes for the number of heads obtained are zero, one, two, three, and four. When a single die is rolled, the outcomes are one, two, three, four, five, and six. These are examples of discrete variables.

A **continuous** variable has an infinite number of values between any two values. Continuous variables are measured. For example, temperature is a continuous variable since the variable can assume any value between 10° and 20° or any other two temperatures or values for that matter. Height and weight are continuous variables. Of course, we are limited by our measuring devices and values of continuous variables are usually "rounded off."

EXAMPLE: Construct a discrete probability distribution for the number of heads when three coins are tossed.

SOLUTION:

Recall that the sample space for tossing three coins is

TTT, TTH, THT, HTT, HHT, HTH, THH, and HHH.

The outcomes can be arranged according to the number of heads, as shown.

0 heads	TTT
1 head	TTH, THT, HTT
2 heads	THH, HTH, HHT
3 heads	HHH

Finally, the outcomes and corresponding probabilities can be written in a table, as shown.

Outcome, x	0	1	2	3
Probability, $P(x)$	$\frac{1}{8}$	$\frac{3}{8}$	$\frac{3}{8}$	$\frac{1}{8}$

The sum of the probabilities of a probability distribution must be 1.

A discrete probability distribution can also be shown graphically by labeling the x axis with the values of the outcomes and letting the values on the y axis represent the probabilities for the outcomes. The graph for the discrete probability distribution of the number of heads occurring when three coins are tossed is shown in Figure 7-1.

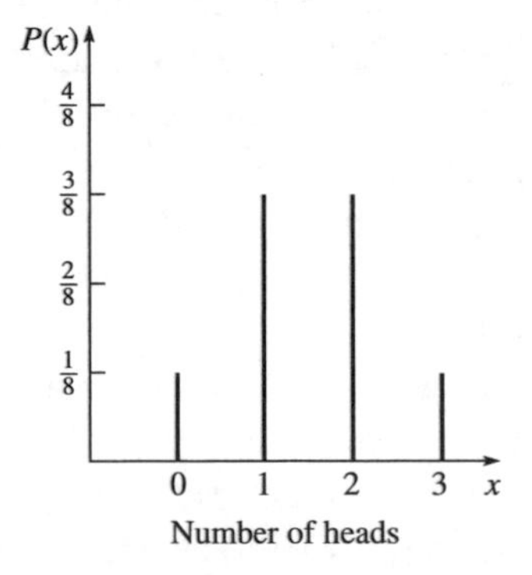

Fig. 7-1.

There are many kinds of discrete probability distributions; however, the distribution of the number of heads when three coins are tossed is a special kind of distribution called a *binomial distribution*.

A binomial distribution is obtained from a probability experiment called a **binomial experiment**. The experiment must satisfy these conditions:

1. Each trial can have only two outcomes or outcomes that can be reduced to two outcomes. The outcomes are usually considered as a success or a failure.
2. There is a fixed number of trials.
3. The outcomes of each trial are independent of each other.
4. The probability of a success must remain the same for each trial.

EXAMPLE: Explain why the probability experiment of tossing three coins is a binomial experiment.

SOLUTION:

In order to be a binomial experiment, the probability experiment must satisfy the four conditions explained previously.

1. There are only two outcomes for each trial, head and tail. Depending on the situation, either heads or tails can be defined as a success and the other as a failure.
2. There is a fixed number of trials. In this case, there are three trials since three coins are tossed or one coin is tossed three times.
3. The outcomes are independent since tossing one coin does not effect the outcome of the other two tosses.
4. The probability of a success (say heads) is $\frac{1}{2}$ and it does not change.

Hence the experiment meets the conditions of a binomial experiment.

Now consider rolling a die. Since there are six outcomes, it cannot be considered a binomial experiment. However, it can be made into a binomial experiment by considering the outcome of getting five spots (for example) a success and every other outcome a failure.

In order to determine the probability of a success for a single trial of a probability experiment, the following formula can be used.

$$_{n}C_{x} \cdot (p)^{x} \cdot (1-p)^{n-x}$$

where $n =$ the total number of trials
$x =$ the number of successes $(1, 2, 3, \ldots, n)$
$p =$ the probability of a success

The formula has three parts: $_{n}C_{x}$ determines the number of ways a success can occur. $(p)^{x}$ is the probability of getting x successes, and $(1-p)^{n-x}$ is the probability of getting $n-x$ failures.

EXAMPLE: A coin is tossed 3 times. Find the probability of getting two heads and a tail in any given order.

SOLUTION:

Since the coin is tossed 3 times, $n=3$. The probability of getting a head (success) is $\frac{1}{2}$, so $p=\frac{1}{2}$ and the probability of getting a tail (failure) is $1-\frac{1}{2}=\frac{1}{2}$; $x=2$ since the problem asks for 2 heads. $(n-x)=3-2=1$.

Hence,

$$
\begin{aligned}
P(2\text{ heads}) &= {}_3C_2 \cdot \left(\frac{1}{2}\right)^2 \left(\frac{1}{2}\right)^1 \\
&= 3 \cdot \left(\frac{1}{4}\right)\left(\frac{1}{2}\right) \\
&= \frac{3}{8}
\end{aligned}
$$

Notice that there were ${}_3C_2$ or 3 ways to get two heads and a tail. The answer $\frac{3}{8}$ is also the same as the answer obtained using classical probability that was shown in the first example in this chapter.

EXAMPLE: A die is rolled 3 times; find the probability of getting exactly one five.

SOLUTION:

Since we are rolling a die 3 times, $n=3$. The probability of getting a 5 is $\frac{1}{6}$. The probability of not getting a 5 is $1-\frac{1}{6}$ or $\frac{5}{6}$. Since a success is getting one five, $x=1$ and $n-x=3-1=2$.

Hence,

$$
\begin{aligned}
P(\text{one }5) &= {}_3C_1 \cdot \left(\frac{1}{6}\right)^1 \cdot \left(\frac{5}{6}\right)^2 \\
&= 3 \cdot \frac{1}{6} \cdot \frac{25}{36} \\
&= \frac{25}{72} \text{ or } 0.3472
\end{aligned}
$$

About 35% of the time, exactly one 5 will occur.

EXAMPLE: An archer hits the bull's eye 80% of the time. If he shoots 5 arrows, find the probability that he will get 4 bull's eyes.

SOLUTION:

$$n=5,\ x=4,\ p=0.8,\ 1-p=1-0.8=0.2$$

$$\begin{aligned} P(4\text{ bull's eyes}) &= {}_5C_4(0.8)^4(0.2)^1 \\ &= 5\cdot 0.08192 \\ &= 0.4096 \end{aligned}$$

In order to construct a probability distribution, the following formula is used:

$${}_nC_x p^x(1-p)^{n-x} \text{ where } x=1,2,3,\ldots n.$$

The next example shows how to use the formula.

EXAMPLE: A die is rolled 3 times. Construct a probability distribution for the number of fives that will occur.

SOLUTION:

In this case, the die is tossed 3 times, so $n=3$. The probability of getting a 5 on a die is $\frac{1}{6}$, and one can get $x=0,1,2$, or 3 fives.

For $x=0$, ${}_3C_0\left(\frac{1}{6}\right)^0\left(\frac{5}{6}\right)^3=0.5787$

For $x=1$, ${}_3C_1\left(\frac{1}{6}\right)^1\left(\frac{5}{6}\right)^2=0.3472$

For $x=2$, ${}_3C_2\left(\frac{1}{6}\right)^2\left(\frac{5}{6}\right)^1=0.0694$

For $x=3$, ${}_3C_3\left(\frac{1}{6}\right)^3\left(\frac{5}{6}\right)^0=0.0046$

Hence, the probability distribution is

Number of fives, x	0	1	2	3
Probability, $P(x)$	0.5787	0.3472	0.0694	0.0046

Note: Most statistics books have tables that can be used to compute probabilities for binomial variables.

PRACTICE

1. A student takes a 5-question true–false quiz. Since the student has not studied, he decides to flip a coin to determine the answers. What is the probability that the student guesses exactly 3 out of 5 correctly?
2. A basketball player makes three-fourths of his free throws. Assume each shot is independent of another shot. Find the probability that he makes the next four free throws.
3. A circuit has 6 breakers. The probability that each breaker will fail is 0.1. If the circuit is activated, find the probability that exactly two breakers will fail. Each breaker is independent of any other breaker.
4. Eight coins are tossed; find the probability of getting exactly 3 heads.
5. A box contains 4 red marbles and 2 white marbles. A marble is drawn and replaced four times. Find the probability of getting exactly 3 red marbles and one white marble.

ANSWERS

1. $n=5,\ x=3,\ p=\frac{1}{2}$

$$\begin{aligned} P(\text{exactly 3 correct}) &= {}_5C_3\left(\frac{1}{2}\right)^3\left(\frac{1}{2}\right)^2 \\ &= (10)\frac{1}{32} \\ &= \frac{5}{16} = 0.3125 \end{aligned}$$

2. $n=4,\ x=4,\ p=\frac{3}{4}$

$$\begin{aligned} P(\text{4 successes}) &= {}_4C_4\left(\frac{3}{4}\right)^4\left(\frac{1}{4}\right)^0 \\ &= \frac{81}{256} \approx 0.3164 \end{aligned}$$

3. $n=6,\ x=2,\ p=(0.1)$

$$\begin{aligned} P(\text{2 will fail}) &= {}_6C_2(0.1)^2(0.9)^4 \\ &= 15(0.006561) = 0.098415 \end{aligned}$$

4. $n=8,\ x=3,\ p=\frac{1}{2}$

$$P(3\text{ heads}) = {}_8C_3\left(\frac{1}{2}\right)^3\left(\frac{1}{2}\right)^5$$
$$= 56\left(\frac{1}{8}\right)\left(\frac{1}{32}\right)$$
$$= 56\cdot\frac{1}{256}$$
$$= \frac{7}{32} = 0.21875$$

5. $n=4,\ x=3,\ p=\frac{2}{3}$

$$P(3\text{ red marbles}) = {}_4C_3\left(\frac{2}{3}\right)^3\left(\frac{1}{3}\right)^1$$
$$= 4\cdot\frac{8}{81}$$
$$= \frac{32}{81} \approx 0.395$$

The Mean and Standard Deviation for a Binomial Distribution

Suppose you roll a die many times and record the number of threes you obtain. Is it possible to predict ahead of time the average number of threes you will obtain? The answer is "Yes." It is called **expected value** or the **mean** of a binomial distribution. This mean can be found by using the formula mean $(\mu) = np$ where n is the number of times the experiment is repeated and p is the probability of a success. The symbol for the mean is the Greek letter μ (mu).

EXAMPLE: A die is tossed 180 times and the number of threes obtained is recorded. Find the mean or expected number of threes.

SOLUTION:

$n = 180$ and $p = \frac{1}{6}$ since there is one chance in 6 to get a three on each roll.

$$\mu = n \cdot p = 180 \cdot \frac{1}{6}$$
$$= 30$$

Hence, one would expect on average 30 threes.

EXAMPLE: Twelve cards are selected from a deck and each card is replaced before the next one is drawn. Find the average number of diamonds.

SOLUTION:

In this case, $n = 12$ and $p = \frac{13}{52}$ or $\frac{1}{4}$ since there are 13 diamonds and a total of 52 cards. The mean is

$$\mu = n \cdot p$$
$$= 12 \cdot \frac{1}{4}$$
$$= 3$$

Hence, on average, we would expect 3 diamonds in the 12 draws.

Statisticians are not only interested in the average of the outcomes of a probability experiment but also in how the results of a probability experiment vary from trial to trial. Suppose we roll a die 180 times and record the number of threes obtained. We know that we would expect to get about 30 threes. Now what if the experiment was repeated again and again? In this case, the number of threes obtained each time would not always be 30 but would vary about the mean of 30. For example, we might get 28 threes one time and 34 threes the next time, etc. How can this variability be explained? Statisticians use a measure called the **standard deviation.** When the standard deviation of a variable is large, the individual values of the variable are spread out from the mean of the distribution. When the standard deviation of a variable is small, the individual values of the variable are close to the mean.

The formula for the standard deviation for a binomial distribution is standard deviation $\sigma = \sqrt{np(1-p)}$. The symbol for the standard deviation is the Greek letter σ (sigma).

EXAMPLE: A die is rolled 180 times. Find the standard deviation of the number of threes.

SOLUTION:

$$n = 180,\ p = \frac{1}{6},\ 1 - p = 1 - \frac{1}{6} = \frac{5}{6}$$

$$\begin{aligned}\sigma &= \sqrt{np(1-p)}\\ &= \sqrt{180 \cdot \frac{1}{6} \cdot \frac{5}{6}}\\ &= \sqrt{25}\\ &= 5\end{aligned}$$

The standard deviation is 5.

Now what does this tell us?

Roughly speaking, most of the values fall within two standard deviations of the mean.

$$\mu - 2\sigma < \text{most values} < \mu + 2\sigma$$

In the die example, we can expect most values will fall between

$$30 - 2 \cdot 5 < \text{most values} < 30 + 2 \cdot 5$$

$$30 - 10 < \text{most values} < 30 + 10$$

$$20 < \text{most values} < 40$$

In this case, if we did the experiment many times we would expect between 20 and 40 threes most of the time. This is an approximate "range of values."

Suppose we rolled a die 180 times and we got only 5 threes, what can be said? It can be said that this is an unusually small number of threes. It can happen by chance, but not very often. We might want to consider some other possibilities. Perhaps the die is loaded or perhaps the die has been manipulated by the person rolling it!

EXAMPLE: An archer hits the bull's eye 80% of the time. If he shoots 100 arrows, find the mean and standard deviation of the number of bull's eyes. If he travels to many tournaments, find the approximate range of values.

SOLUTION:

$$n = 100,\ p = 0.80,\ 1 - p = 1 = 0.80 = 0.20$$

$$\text{mean: } \mu = np$$
$$= 100(0.80)$$
$$= 80$$

$$\text{standard deviation: } \sqrt{np(1-p)}$$
$$= \sqrt{100(0.8)(0.2)}$$
$$= \sqrt{16}$$
$$= 4$$

Approximate range of values:

$$\mu - 2\sigma < \text{most values} < \mu + 2 \cdot (\sigma)$$
$$80 - 2 \cdot (4) < \text{most values} < 80 + 2 \cdot (4)$$
$$72 < \text{most values} < 88$$

Hence, most of his scores will be somewhere between 72 and 88.

Note: The concept of the standard deviation is much more complex than what is presented here. Additional information on the standard deviation will be presented in Chapter 9. More information on the standard deviation can also be found in all statistics textbooks.

PRACTICE

1. Twenty cards are selected from a deck of 52 cards. Each card is replaced before the next card is selected. Find the mean and standard deviation of the number of clubs selected.
2. A coin is tossed 1000 times. Find the mean and standard deviation of the number of heads that will occur.
3. A 50-question multiple choice exam is given. There are four choices for each question. Find the mean and standard deviation of the number of correct answers a student will get if he selects each answer at random.
4. A die is rolled 720 times. Find the mean, standard deviation, and approximate range of values for the number of threes obtained.
5. A factory manufactures microchips of which 4% are defective. Find the average number of defective microchips in a lot of 500. Also, find the standard deviation and approximate range of values.

ANSWERS

1. $n=20$, $p=\frac{1}{4}$, $1-p=1-\frac{1}{4}=\frac{3}{4}$

$$\mu = n \cdot p$$
$$= 20 \cdot \frac{1}{4}$$
$$= 5$$

$$\sigma = \sqrt{np(1-p)}$$
$$= \sqrt{20 \cdot \frac{1}{4} \cdot \frac{3}{4}}$$
$$= \sqrt{3.75} \approx 1.936$$

2. $n=1000$, $p=\frac{1}{2}$, $1-p=1-\frac{1}{2}=\frac{1}{2}$

$$\mu = n \cdot p$$
$$= 1000 \cdot \frac{1}{2}$$
$$= 500$$

$$\sigma = \sqrt{np(1-p)}$$
$$= \sqrt{1000 \cdot \frac{1}{2} \cdot \frac{1}{2}} = \sqrt{250} \approx 15.81$$

3. $n=50$, $p=\frac{1}{4}$, $1-p=1-\frac{1}{4}=\frac{3}{4}$

$$\sigma = n \cdot p$$
$$= 50 \cdot \frac{1}{4}$$
$$= 12.5$$

$$\sigma = \sqrt{np(1-p)}$$
$$= \sqrt{50 \cdot \frac{1}{4} \cdot \frac{3}{4}}$$
$$= \sqrt{9.375} \approx 3.06$$

4. $n = 720,\ p = \frac{1}{6},\ 1 - p = 1 - \frac{1}{6} = \frac{5}{6}$

$$\begin{aligned}\mu &= n \cdot p \\ &= 720\left(\frac{1}{6}\right) \\ &= 120\end{aligned}$$

$$\begin{aligned}\sigma &= \sqrt{np(1-p)} \\ &= \sqrt{720\left(\frac{1}{6}\right)\left(\frac{5}{6}\right)} \\ &= \sqrt{100} = 10\end{aligned}$$

$$\mu - 2\sigma < \text{most scores} < \mu + 2\sigma$$
$$120 - 2 \cdot 10 < \text{most scores} < 120 + 2 \cdot 10$$
$$100 < \text{most scores} < 140$$

5. $n = 500,\ p = 0.04,\ 1 - p = 1 - 0.04 = 0.96$

$$\begin{aligned}\mu &= n \cdot p \\ &= 500(0.04) \\ &= 20\end{aligned}$$

$$\begin{aligned}\sigma &= \sqrt{500(0.04)(0.96)} \\ &= \sqrt{19.2} \approx 4.38\end{aligned}$$

$$\mu - 2\sigma < \text{most values} < \mu + 2\sigma$$
$$20 - 2(4.38) < \text{most values} < 20 + 2(4.38)$$
$$11.24 < \text{most values} < 28.76$$

CHAPTER QUIZ

1. How many outcomes are there for a binomial experiment?
   a. 0
   b. 1
   c. 2
   d. It varies

2. The sum of the probabilities of all outcomes in a probability distribution is

   a. 0
   b. 1
   c. 2
   d. It varies

3. Which one is not a requirement of a binomial experiment?

   a. There are 2 outcomes for each trial.
   b. There is a fixed number of trials.
   c. The outcomes must be dependent.
   d. The probability of a success must be the same for all trials.

4. The formula for the mean of a binomial distribution is

   a. np
   b. $np(1-p)$
   c. $n(1-p)$
   d. $\sqrt{np(1-p)}$

5. The formula for a standard deviation of a binomial distribution is

   a. np
   b. $np(1-p)$
   c. $n(1-p)$
   d. $\sqrt{np(1-p)}$

6. If 30% of commuters ride to work on a bus, find the probability that if 8 workers are selected at random, 3 will ride the bus.

   a. 0.361
   b. 0.482
   c. 0.254
   d. 0.323

7. If 10% of the people who take a certain medication get a headache, find the probability that if 5 people take the medication, one will get a headache.

   a. 0.328
   b. 0.136
   c. 0.472
   d. 0.215

8. A survey found that 30% of all Americans have eaten pizza for breakfast. If 500 people are selected at random, the mean number of people who have eaten pizza for breakfast is

   a. 100
   b. 150
   c. 200
   d. 230

9. A survey found that 10% of older Americans have given up driving. If a sample of 1000 Americans is taken, the standard deviation of the sample is

   a. 10
   b. 8.42
   c. 9.49
   d. 5

10. A survey found that 50% of adults get the daily news from radio. If a sample of 64 adults is selected, the approximate range of the number of people who get their news from the radio is

    a. 24 and 40
    b. 30 and 34
    c. 28 and 32
    d. 26 and 36

Probability Sidelight

PASCAL'S TRIANGLE

Blaise Pascal (1623–1662) was a French mathematician and philosopher. He made many contributions to mathematics in areas of number theory, geometry, and probability. He is credited along with Fermat for the beginnings of the formal study of probability. He is given credit for developing a triangular array of numbers known as Pascal's triangle, shown here.

1

1 2 1

1 3 3 1

1 4 6 4 1

1 5 10 10 5 1

1 6 15 20 15 6 1

Each number in the triangle is the sum of the number above and to the right of it and the number above and to the left of it. For example, the number 10 in the fifth row is found by adding the 4 and 6 in the fourth row. The number 15 in the sixth row is found by adding the 5 and 10 in the previous row.

The numbers in each row represent the number of different outcomes when coins are tossed. For example, the numbers in row 3 are 1, 3, 3, and 1. When 3 coins are tossed, the outcomes are

THH HTT

HTH THT

HHH HHT TTH TTT

Notice that there is only one way to get 3 heads. There are 3 different ways to get 2 heads and a tail. There are 3 different ways to get two tails and a head and there is one way to get 3 tails. The same results apply to the genders of the children in a family with three children.

Another property of the triangle is that it represents the answer to the number of combination of n items taking r items at a time as shown.

$_0C_0$

$_1C_0$ $_1C_1$

$_2C_0$ $_2C_1$ $_2C_2$

$_3C_0$ $_3C_1$ $_3C_2$ $_3C_3$

etc.

The numbers in the triangle have applications in other areas of mathematics such as algebra and graph theory.

It is interesting to note that Pascal included his triangle in a book he wrote in 1653. It wasn't printed until 1665. It is not known if Pascal developed the triangle on his own or heard about it from someone else; however, a similar version of the triangle was found in a Chinese manuscript written by Chi Shi-Kie in 1303!

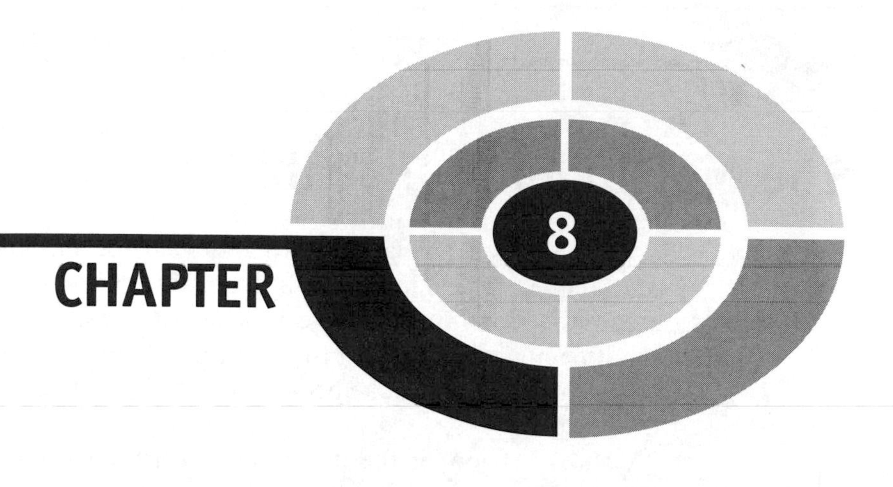

Other Probability Distributions

Introduction

The last chapter explained the concepts of the binomial distribution. There are many other types of commonly used discrete distributions. A few are the multinomial distribution, the hypergeometric distribution, the Poisson distribution, and the geometric distribution. This chapter briefly explains the basic concepts of these distributions.

The Multinomial Distribution

Recall that for a probability experiment to be binomial, two outcomes are necessary. But if each trial of a probability experiment has more than two outcomes, a distribution that can be used to describe the experiment is called a **multinomial distribution**. In addition, there must be a fixed number of independent trials, and the probability for each success must remain the same for each trial.

A short version of the multinomial formula for three outcomes is given next. If X consists of events E_1, E_2, and E_3, which have corresponding probabilities of p_1, p_2, and p_3 of occurring, where x_1 is the number of times E_1 will occur, x_2 is the number of times E_2 will occur, and x_3 is the number of times E_3 will occur, then the probability of X is

$$\frac{n!}{x_1!x_2!x_3!} \cdot p_1^{x_1} \cdot p_2^{x_2} \cdot p_3^{x_3} \text{ where } x_1 + x_2 + x_3 = n \text{ and } p_1 + p_2 + p_3 = 1.$$

EXAMPLE: In a large city, 60% of the workers drive to work, 30% take the bus, and 10% take the train. If 5 workers are selected at random, find the probability that 2 will drive, 2 will take the bus, and 1 will take the train.

SOLUTION:

$n=5$, $x_1=2$, $x_2=2$, $x_3=1$ and $p_1=0.6$, $p_2=0.3$, and $p_3=0.1$

Hence, the probability that 2 workers will drive, 2 will take the bus, and one will take the train is

$$\frac{5!}{2!2!1!} \cdot (0.6)^2(0.3)^2(0.1)^1 = 30 \cdot (0.36)(0.09)(0.1) = 0.0972$$

EXAMPLE: A box contains 5 red balls, 3 blue balls, and 2 white balls. If 4 balls are selected with replacement, find the probability of getting 2 red balls, one blue ball, and one white ball.

SOLUTION:

$n=4$, $x_1=2$, $x_2=1$, $x_3=1$, and $p_1=\frac{5}{10}$, $p_2=\frac{3}{10}$, and $p_3=\frac{2}{10}$. Hence, the probability of getting 2 red balls, one blue ball, and one white ball is

$$\frac{4!}{2!1!1!}\left(\frac{5}{10}\right)^2\left(\frac{3}{10}\right)^1\left(\frac{2}{10}\right)^1 = 12\left(\frac{3}{200}\right) = \frac{9}{50} = 0.18$$

PRACTICE

1. At a swimming pool snack bar, the probabilities that a person buys one item, two items, or three items are 0.3, 0.4, and 0.3. If six people are selected at random, find the probability that 2 will buy one item, 3 will buy two items, and 1 will buy three items.
2. A survey of adults who go out once a week showed 60% choose a movie, 30% choose dinner and a play, and 10% go shopping. If 10 people are selected, find the probability that 5 will go to a movie, 4 will go to dinner and a play, and one will go shopping.
3. A box contains 5 white marbles, 3 red marbles, and 2 green marbles. If 5 marbles are selected with replacement, find the probability that 2 will be white, 2 will be red, and one will be green.
4. Automobiles are randomly inspected in a certain state. The probabilities for having no violations, 1 violation, and 2 or more violations are 0.50, 0.30, and 0.20 respectively. If 10 automobiles are inspected, find the probability that 5 will have no violations, 3 will have one violation, and 2 will have 2 or more violations.
5. According to Mendel's theory, if tall and colorful plants are crossed with short and colorless plants, the corresponding probabilities are $\frac{9}{16}$, $\frac{3}{16}$, $\frac{3}{16}$, and $\frac{1}{16}$ for tall and colorful, tall and colorless, short and colorful, and short and colorless. If 8 plants are selected, find the probability that 3 will be tall and colorful, 2 will be tall and colorless, 2 will be short and colorful and 1 will be short and colorless.

ANSWERS

1. $n=6$, $x_1=2$, $x_2=3$, $x_3=1$, and $p_1=0.3$, $p_2=0.4$, and $p_3=0.3$

   The probability is $\frac{6!}{2!3!1!} \cdot (0.3)^2(0.4)^3(0.3)^1 = 60(0.001728) = 0.10368$

2. $n=10$, $x_1=5$, $x_2=4$, $x_3=1$, and $p_1=0.6$, $p_2=0.3$, and $p_3=0.1$

   The probability is $\frac{10!}{5!4!1!}(0.6)^5(0.3)^4(0.1)^1 = 1260(0.000062986) =$ 0.07936

3. $n=5$, $x_1=2$, $x_2=2$, $x_3=1$, and $p_1=\frac{5}{10}$, $p_2=\frac{3}{10}$, and $p_3=\frac{2}{10}$

   The probability is $\frac{5!}{2!2!1!}\left(\frac{5}{10}\right)^2\left(\frac{3}{10}\right)^2\left(\frac{2}{10}\right)^1=30(0.0045)=0.135$

4. $n=10$, $x_1=5$, $x_2=3$, $x_3=2$, and $p_1=0.5$, $p_2=0.3$, and $p_3=0.2$

   The probability is $\frac{10!}{5!3!2!}(0.5)^5(0.3)^3(0.2)^2=2520\,(0.00003375)=$ 0.08505

5. $n=8$, $x_1=3$, $x_2=2$, $x_3=2$, $x_4=1$, and $p_1=\frac{9}{16}$, $p_2=\frac{3}{16}$, $p_3=\frac{3}{16}$, and $p_4=\frac{1}{16}$

   The probability is $\frac{8!}{3!2!2!1!}\left(\frac{9}{16}\right)^3\left(\frac{3}{16}\right)^2\left(\frac{3}{16}\right)^2\left(\frac{1}{16}\right)^1=1680(0.000013748)=$ 0.0231

The Hypergeometric Distribution

When a probability experiment has two outcomes and the items are selected without replacement, the hypergeometric distribution can be used to compute the probabilities. When there are two groups of items such that there are a items in the first group and b items in the second group, so that the total number of items is $a+b$, the probability of selecting x items from the first group and $n-x$ items from the second group is

$$\frac{{}_aC_x \cdot {}_bC_{n-x}}{{}_{a+b}C_n}$$

where n is the total number of items selected without replacement.

EXAMPLE: A committee of 4 people is selected at random without replacement from a group of 6 men and 4 women. Find the probability that the committee consists of 2 men and 2 women.

SOLUTION:

Since there are 6 men and 2 women, $a=6$, $b=4$ and $n=6+4$ or 10. Since the committee consists of 2 men and 2 women, $x=2$, and $n-x=4-2=2$.

The probability is

$$\frac{{}_6C_2 \cdot {}_4C_2}{{}_{10}C_4} = \frac{\dfrac{6!}{2!4!} \cdot \dfrac{4!}{2!2!}}{\dfrac{10!}{4!6!}} = \frac{15 \cdot 6}{210} = \frac{90}{210} = \frac{3}{7} \approx 0.429$$

EXAMPLE: A lot of 12 oxygen tanks contains 3 defective ones. If 4 tanks are randomly selected and tested, find the probability that exactly one will be defective.

SOLUTION:

Since there are 3 defective tanks and 9 good tanks, $a = 3$ and $b = 9$. If 4 tanks are randomly selected and we want to know the probability that exactly one is defective, $n = 4$, $x = 1$, and $n - x = 4 - 1 = 3$. The probability then is

$$\frac{{}_3C_1 \cdot {}_9C_3}{{}_{12}C_4} = \frac{\dfrac{3!}{1!2!} \cdot \dfrac{9!}{3!6!}}{\dfrac{12!}{4!8!}} = \frac{3 \cdot 84}{495} \approx 0.509$$

PRACTICE

1. In a box of 12 shirts there are 5 defective ones. If 5 shirts are sold at random, find the probability that exactly two are defective.
2. In a fitness club of 18 members, 10 prefer the exercise bicycle and 8 prefer the aerobic stepper. If 6 members are selected at random, find the probability that exactly 3 use the bicycle.
3. In a shipment of 10 lawn chairs, 6 are brown and 4 are blue. If 3 chairs are sold at random, find the probability that all are brown.
4. A class consists of 5 women and 4 men. If a committee of 3 people is selected at random, find the probability that all 3 are women.
5. A box contains 3 red balls and 3 white balls. If two balls are selected at random, find the probability that both are red.

ANSWERS

1. $a = 5,\ b = 7,\ n = 5,\ x = 2,\ n - x = 5 - 2 = 3$

   The probability is $\dfrac{{}_5C_2 \cdot {}_7C_3}{{}_{12}C_5} = \dfrac{10 \cdot 35}{792} \approx 0.442$

2. $a = 10,\ b = 8,\ n = 6,\ x = 3,\ n - x = 3$

   The probability is $\dfrac{{}_{10}C_3 \cdot {}_8C_3}{{}_{18}C_6} = \dfrac{120 \cdot 56}{18,564} \approx 0.362$

3. $a = 6,\ b = 4,\ n = 3,\ x = 3,\ n - x = 3 - 3 = 0$

   The probability is $\dfrac{{}_6C_3 \cdot {}_4C_0}{{}_{10}C_3} = \dfrac{20 \cdot 1}{120} = 0.167$

4. $a = 5,\ b = 4,\ n = 3,\ x = 3,\ n - x = 3 - 3 = 0$

   The probability is $\dfrac{{}_5C_3 \cdot {}_4C_0}{{}_9C_3} = \dfrac{10 \cdot 1}{84} \approx 0.119$

5. $a = 3,\ b = 3,\ n = 2,\ x = 2,\ n - x = 2 - 2 = 0$

   The probability is $\dfrac{{}_3C_2 \cdot {}_3C_0}{{}_6C_2} = \dfrac{3 \cdot 1}{15} = 0.2$

The Geometric Distribution

Suppose you flip a coin several times. What is the probability that the first head appears on the third toss? In order to answer this question and other similar probability questions, the geometric distribution can be used. The formula for the probability that the first success occurs on the nth trial is

$$(1 - p)^{n-1}p$$

where p is the probability of a success and n is the trial number of the first success.

EXAMPLE: A coin is tossed. Find the probability that the first head occurs on the third toss.

SOLUTION:

The outcome is TTH; hence, $n = 3$ and $p = \frac{1}{2}$, so the probability of getting two tails and then a head is $\frac{1}{2} \cdot \frac{1}{2} \cdot \frac{1}{2} = \frac{1}{8}$.

Using the formula given above, $\left(1-\frac{1}{2}\right)^{3-1} \cdot \frac{1}{2} = \left(\frac{1}{2}\right)^2 \left(\frac{1}{2}\right) = \frac{1}{8}$

EXAMPLE: A die is rolled. Find the probability of getting the first three on the fourth roll.

SOLUTION:

Let $p = \frac{1}{6}$ and $n = 4$; hence, $\left(1-\frac{1}{6}\right)^{4-1} \frac{1}{6} = \left(\frac{5}{6}\right)^3 \left(\frac{1}{6}\right) = \frac{125}{1296} \approx 0.096$

The geometric distribution can be used to answer the question, "How long on average will I have to wait for a success?"

Suppose a person rolls a die until a five is obtained. The five could occur on the first roll (if one is lucky), on the second roll, on the third roll, etc. Now the question is, "On average, how many rolls would it take to get the first five?" The answer is that if the probability of a success is p, then the average or expected number of independent trials it would take to get a success is $\frac{1}{p}$. In the dice situation, it would take on average $1 \div \frac{1}{6}$ or 6 trials to get a five. This is not so hard to believe since a five would occur on average one time in every six rolls because the probability of getting a five is $\frac{1}{6}$.

EXAMPLE: A coin is tossed until a head is obtained. On average, how many trials would it take?

SOLUTION:

Since the probability of getting a head is $\frac{1}{2}$, it would take $1 \div p$ trials.

$$1 \div \frac{1}{2} = 1 \cdot \frac{2}{1} = 2$$

On average it would take two trials.

Now suppose we ask, "On average, how many trials would it take to get two fives?" In this case, one five would occur on average once in the next six trials, so the second five would occur on average once in the next six trials. In general we would expect k successes on average in k/p trials.

EXAMPLE: If cards are selected from a deck and replaced, how many trials would it take on average to get two clubs?

SOLUTION:

Since there are 13 clubs in a deck of 52 cards, $P(\text{club}) = \frac{13}{52} = \frac{1}{4}$. The expected number of trials for selecting two clubs would be $\frac{2}{\frac{1}{4}}$ or $2 \div \frac{1}{4} = 2 \cdot 4 = 8$ trials.

This type of problem uses what is called the negative binomial distribution, which is a generalization of the geometric distribution.

Another interesting question one might ask is, "On average how many rolls of a die would it take to get all the faces, one through six, on a die?" In this case, the first roll would give one of the necessary numbers, so the probability of getting a number needed on the first roll would be one. On the second roll, the probability of getting a number needed would be $\frac{5}{6}$ since there are 5 remaining needed numbers. The average number of rolls would be $(\frac{1}{5/6})$ or $\frac{6}{5}$. Since two numbers have been obtained, the probability of getting the next number would be $\frac{4}{6}$. The average number of rolls would be $(\frac{1}{4/6})$ or $\frac{6}{4}$. This would continue until all numbers are obtained. So the average number of rolls it would take to get all the numbers, one through six, would be $1 + \frac{6}{5} + \frac{6}{4} + \frac{6}{3} + \frac{6}{2} + \frac{6}{1} = 14.7$. Hence on average it would take about 14.7 rolls to get all the numbers one through six.

EXAMPLE: A children's cereal manufacturer packages one toy space craft in each box. If there are 4 different toys, and they are equally distributed, find the average number of boxes a child would have to purchase to get all four.

SOLUTION:

The probabilities are $1, \frac{3}{4}, \frac{2}{4}$, and $\frac{1}{4}$. The average number of boxes for each are $\frac{1}{1}, \frac{1}{(3/4)}, \frac{1}{(2/4)}$, and $\frac{1}{(1/4)}$ so the total is $1 + \frac{4}{3} + \frac{4}{2} + \frac{4}{1} = 8\frac{1}{3}$ which would mean a child on average would need to purchase 9 boxes of cereal since he or she could not buy $\frac{1}{3}$ of a box.

PRACTICE

1. A card from an ordinary deck of cards is selected and then replaced. Another card is selected, etc. Find the probability that the first club will occur on the third draw.
2. A die is tossed until a one or a two is obtained. Find the expected number of tosses.
3. On average how many rolls of a die will it take to get 3 fours?

4. A coin is tossed until 4 heads are obtained. What is the expected number of tosses?
5. A service station operator gives a scratch-off card with each fill up over 8 gallons. On each card is one of 5 colors. When a customer gets all five colors, he wins 10 gallons of gasoline. Find the average number of fill ups needed to win the 10 gallons.

ANSWERS

1. $\left(\frac{3}{4}\right)\left(\frac{3}{4}\right)\left(\frac{1}{4}\right) = \frac{9}{64}$
2. $p = \frac{2}{6} = \frac{1}{3}; \frac{1}{p} = \frac{1}{\left(\frac{1}{3}\right)} = 1 \div \frac{1}{3} = 1 \cdot 3 = 3$
3. $\frac{3}{\left(\frac{1}{6}\right)} = 3 \div \frac{1}{6} = 3 \cdot 6 = 18$
4. $\frac{4}{\frac{1}{2}} = 4 \div \frac{1}{2} = 4 \cdot 2 = 8$
5. $1 + \frac{5}{4} + \frac{5}{3} + \frac{5}{2} + \frac{5}{1} = 11\frac{5}{12}$, i.e. 12 fill ups

The Poisson Distribution

Another commonly used discrete distribution is the Poisson distribution (named after Simeon D. Poisson, 1781–1840). This distribution is used when the variable occurs over a period of time, volume, area, etc. For example, it can be used to describe the arrivals of airplanes at an airport, the number of phone calls per hour for a 911 operator, the density of a certain species of plants over a geographic region, or the number of white blood cells on a fixed area.

The probability of x successes is

$$\frac{e^{-\lambda}\lambda^x}{x!}$$

where e is a mathematical constant ≈ 2.7183 and λ is the mean or expected value of the variable.

Note: The computations require a scientific calculator. Also, tables for values used in the Poisson distribution are available in some statistics textbooks.

EXAMPLE: If there are 150 typographical errors randomly distributed in a 600-page manuscript, find the probability that any given page has exactly two errors.

SOLUTION:

Find the mean numbers of errors: $\lambda = \frac{150}{600} = \frac{1}{4}$ or 0.25. In other words, there is an average of 0.25 errors per page. In this case, $x = 2$, so the probability of selecting a page with exactly two errors is

$$\frac{e^{-\lambda}\lambda^x}{x!} = \frac{(2.7183)^{-0.25} \cdot (0.25)^2}{2!} = 0.024$$

Hence the probability of two errors is about 2.4%.

EXAMPLE: A hotline with a toll-free number receives an average of 4 calls per hour. For any given hour, find the probability that it will receive exactly 6 calls.

SOLUTION:

The mean $\lambda = 4$ and $x = 6$. The probability is

$$\frac{e^{-\lambda}\lambda^x}{x!} = \frac{(2.7183)^{-4} \cdot (4)^6}{6!} = 0.104$$

Hence there is about a 10.4% chance that the hotline will receive 6 calls.

EXAMPLE: A videotape has an average of two defects for every 1000 feet. Find the probability that in a length of 2000 feet, there are 5 defects.

SOLUTION:

If there are 2 defects per 1000 feet, then the mean number of defects for 2000 feet would be $2 \cdot 2 = 4$. In this case, $x = 5$. The probability then is

$$\frac{e^{-\lambda}\lambda^x}{x!} = \frac{(2.7183)^{-4} \cdot 4^5}{5!} = 0.156$$

PRACTICE

1. A telemarketing company gets on average 6 orders per 1000 calls. If a company calls 500 people, find the probability of getting 2 orders.
2. A crime study for a geographic area showed an average of one home invasion per 40,000 homes. If an area contains 60,000 homes, find the probability of exactly 3 home invasions.
3. The average number of phone inquiries to a toll-free number for a computer help line is 6 per hour. Find the probability that for a specific hour, the company receives 10 calls.
4. A company receives on average 9 calls every time it airs its commercial. Find the probability of getting 20 calls if the commercial is aired four times a day.
5. A trucking firm experiences breakdowns for its trucks on the average of 3 per week. Find the probability that for a given week 5 trucks will experience breakdowns.

ANSWERS

1. The average is 6 orders per 1000 calls, so the average for 500 calls would be 3 orders. $\lambda = 3$ and $x = 2$

$$\frac{e^{-\lambda}\lambda^x}{x!} = \frac{(2.7183)^{-3} \cdot 3^2}{2!} \approx 0.224$$

2. An average of one home invasion for 40,000 homes means that for 60,000 homes, the average would be 1.5 since the ratio $\frac{60,000}{40,000} = 1.5$ and $x = 3$.

$$\frac{e^{-\lambda}\lambda^x}{x!} = \frac{(2.7183)^{-1.5} \cdot (1.5)^3}{3!} \approx 0.126$$

3. $\lambda = 6$, $x = 10$

$$\frac{e^{-\lambda}\lambda^x}{x!} = \frac{(2.7183)^{-6} \cdot 6^{10}}{10!} \approx 0.041$$

4. $\lambda = 4 \cdot 9 = 36$ and $x = 20$

$$\frac{e^{-\lambda}\lambda^x}{x!} = \frac{(2.1783)^{-36} \cdot 36^{20}}{20!} \approx 0.0013$$

5. $\lambda = 3$ and $x = 5$

$$\frac{e^{-\lambda}\lambda^x}{x!} = \frac{(2.7183)^{-3} \cdot 3^5}{5!} \approx 0.101$$

Summary

There are many types of discrete probability distributions besides the binomial distribution. The most common ones are the multinomial distribution, the hypergeometric distribution, the geometric distribution, and the Poisson distribution.

The multinomial distribution is an extension of the binomial distribution and is used when there are three or more independent outcomes for a probability experiment.

The hypergeometric distribution is used when sampling is done without replacement. The geometric distribution is used to determine the probability of an outcome occurring on a specific trial. It can also be used to find the probability of the first occurrence of an outcome.

The Poisson distribution is used when the variable occurs over a period of time, over a period of area or volume, etc.

In addition there are other discrete probability distributions used in mathematics; however, these are beyond the scope of this book.

CHAPTER QUIZ

1. Which distribution requires that sampling be done without replacement?

   a. Geometric
   b. Multinomial
   c. Hypergeometric
   d. Poisson

2. Which distribution can be used when there are 3 or more outcomes?

   a. Geometric
   b. Multinomial
   c. Hypergeometric
   d. Poisson

3. Which distribution can be used when the variable occurs over time?

   a. Geometric
   b. Multinomial
   c. Hypergeometric
   d. Poisson

4. Which distribution can be used to determine the probability of an outcome occurring on a specific trial?

   a. Geometric
   b. Multinomial
   c. Hypergeometric
   d. Poisson

5. The probabilities that a page of a training manual will have 0, 1, 2, or 3 typographical errors are 0.75, 0.15, 0.10, and 0.05 respectively. If 6 pages are randomly selected, the probability that 2 will contain no errors, 2 will contain one error, one will contain 2 errors, and one will contain 3 errors is

   a. 0.078
   b. 0.063
   c. 0.042
   d. 0.011

6. A die is rolled four times. The probability of getting two 3s, one 4, and one 5 is

   a. $\frac{1}{36}$
   b. $\frac{1}{72}$
   c. $\frac{1}{108}$
   d. $\frac{5}{36}$

7. If 5 cards are drawn from a deck without replacement, the probability that exactly three clubs will be selected is

   a. 0.08
   b. 0.02
   c. 0.16
   d. 0.003

8. Of the 20 automobiles in a used car lot, 8 are white. If 6 are selected at random to be sold at an auction, the probability that exactly 2 are white is

   a. 0.267
   b. 0.482
   c. 0.511
   d. 0.358

9. A board of directors consists of 7 women and 5 men. If a slate of 4 officers is selected at random, the probability that exactly 2 officers are men is

   a. 0.375
   b. 0.424
   c. 0.261
   d. 0.388

10. The number of boating accidents on a large lake follows a Poisson distribution. The probability of an accident is 0.01. If there are 500 boats on the lake on a summer day, the probability that there will be exactly 4 accidents will be

    a. 0.192
    b. 0.263
    c. 0.175
    d. 0.082

11. About 5% of the population carries a genetic trait. Assume the distribution is Poisson. The probability that in a group of 100 randomly selected people 7 people carry the gene is

    a. 0.256
    b. 0.104
    c. 0.309
    d. 0.172

12. When a coin is tossed, the probability of getting the first tail on the fourth toss is

    a. $\frac{1}{2}$
    b. $\frac{3}{4}$

c. $\frac{1}{8}$

d. $\frac{1}{16}$

13. An eight-sided die is rolled. The average number of tosses that it will take to get a 6 is

    a. 6
    b. 16
    c. 8
    d. 12

14. A ten-sided die is rolled; the average number of tosses that it will take to get three 3s is

    a. 30
    b. 8
    c. 18
    d. 40

15. On average how many rolls of an eight-sided die will it take to get all the faces at least once?

    a. 28.62
    b. 64
    c. 21.74
    d. 32

Probability Sidelight

SOME THOUGHTS ABOUT LOTTERIES

Today we are bombarded with lotteries. Almost everyone has the fantasy of winning mega millions for a buck. Each month the prizes seem to be getting larger and larger. Each type of lottery gives the odds and the amount of the winnings. For some lotteries, the amount you win is based on the number of people who play. However, the more people who play, the more chance there will be of multiple winners.

The odds and the expected value of a lottery game can be computed using combinations and the probability rules. For example, a lottery game in Pennsylvania is called "Match 6 Lotto." For this game, a player selects

6 numbers from the numbers 1 to 49. If the player matches all six numbers, the player wins a minimum of \$500,000. (Note: There are other ways of winning and if there is no winner, the prize money is held over until the next drawing and increased a certain amount by the number of new players.) For now, just winning the \$500,000 will be considered. In order to figure the odds, it is necessary to figure the number of winning combinations. In this case, we are selecting without regard to order 6 numbers from 49 numbers. Hence there are ${}_{49}C_6$ or 13,983,816 ways to select a ticket. However, the odds given in the lottery brochure are 1 : 4,661,272. The reason is because if you select 6 numbers, you can have the computer select two more sets of six numbers, giving you three chances to win. So dividing 13,983.86 by 3, you get 4,661,272, and the odds are 1 : 4,661,272.

Now let's make some sense of this. There are 60 seconds in a minute, 60 minutes in an hour and 24 hours in a day. So there are $60 \times 60 \times 24 =$ 86,400 seconds in a day. If you divide 4,661,272 by 86,400, you get approximately 54 days. So selecting a winner would be like selecting a given random second in a time period of 54 days!

What about a guaranteed method to win the lottery? It does exist. If you purchased all possible number combinations, then you would be assured of winning, wouldn't you? But is it possible?

A group of investors from Melbourne, Australia, decided to try. At the time, they decided to attempt to win a \$27 million prize given by the Virginia State Lottery. The lottery consisted of selecting 6 numbers out of a possible 44. This means that they would have to purchase ${}_{44}C_6$ or 7,059,052 different tickets. At \$1 a ticket, they would need to raise \$7,059,052. However, the profit would be somewhere near \$20 million if they won. Next there is always a possibility of having to split the winnings with other winners, thus reducing the profit. Finally, they need to buy the 7 or so million tickets within the 72 hour time frame. This group sent out teams, and they were able to purchase about 5 million tickets. They did win the money without having to split the profit! Some objections were raised by the other players (i.e., losers), but the group was eventually paid off.

By the way, it does not matter which numbers you play on the lottery since the drawing is random and every combination has the same probability of occurring. Some people suggest that unusual combinations such as 1, 2, 3, 4, 5, and 6 are better, since there is less of a chance of having to split your winnings if you do indeed win.

So what does this all mean? I heard a mathematician sum it all up by saying that you have the same chance of winning a big jackpot on a state lottery, whether or not you purchase a ticket.

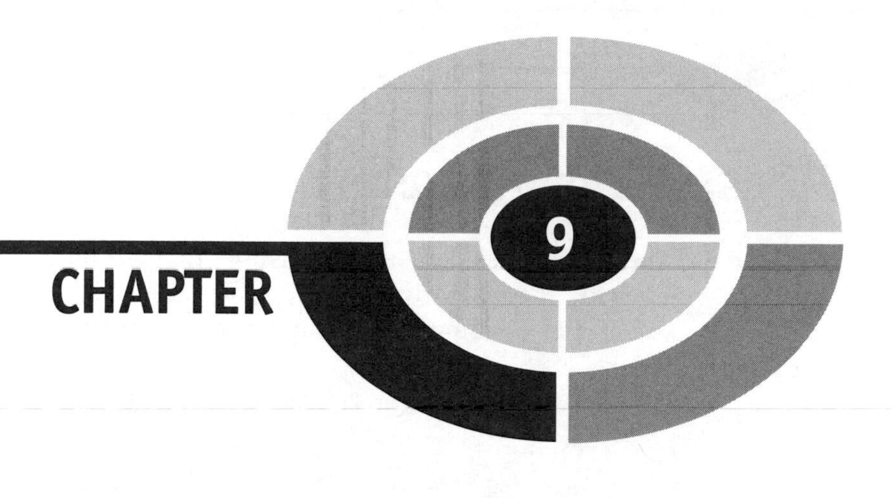

The Normal Distribution

Introduction

A branch of mathematics that uses probability is called *statistics*. **Statistics** is the branch of mathematics that uses observations and measurements called data to analyze, summarize, make inferences, and draw conclusions based on the data gathered. This chapter will explain some basic concepts of statistics such as measures of average and measures of variation. Finally, the relationship between probability and normal distribution will be explained in the last two sections.

Measures of Average

There are three statistical measures that are commonly used for average. They are the *mean, median*, and *mode*. The **mean** is found by adding the data values and dividing by the number of values.

EXAMPLE: Find the mean of 18, 24, 16, 15, and 12.

SOLUTION:

Add the values: $18 + 24 + 16 + 15 + 12 = 85$
Divide by the number of values, 5: $85 \div 5 = 17$
Hence the mean is 17.

EXAMPLE: The ages of 6 executives are 48, 56, 42, 52, 53 and 52. Find the mean.

SOLUTION:

Add: $48 + 56 + 42 + 52 + 53 + 52 = 303$
Divide by 6: $303 \div 6 = 50.5$
Hence the mean age is 50.5.

The **median** is the middle data value if there is an odd number of data values or the number halfway between the two data values at the center, if there is an even number of data values, when the data values are arranged in order.

EXAMPLE: Find the median of 18, 24, 16, 15, and 12.

SOLUTION:

Arrange the data in order: 12, 15, 16, 18, 24
Find the middle value: 12, 15, $\underline{16}$, 18, 24
The median is 16.

EXAMPLE: Find the median of the number of minutes 10 people had to wait in a checkout line at a local supermarket: 3, 0, 8, 2, 5, 6, 1, 4, 1, and 0.

SOLUTION:

Arrange the data in order: 0, 0, 1, 1, 2, 3, 4, 5, 6, 8
The middle falls between 2 and 3; hence, the median is $(2 + 3) \div 2 = 2.5$.

The third measure of average is called the *mode*. The **mode** is the data value that occurs most frequently.

EXAMPLE: Find the mode for 22, 27, 30, 42, 16, 30, and 18.

SOLUTION:

Since 30 occurs twice and more frequently than any other value, the mode is 30.

EXAMPLE: Find the mode for 2, 3, 3, 3, 4, 4, 6, 6, 6, 8, 9, and 10.

SOLUTION:

In this example, 3 and 6 occur most often; hence, 3 and 6 are used as the mode. In this case, we say that the distribution is *bimodal.*

EXAMPLE: Find the mode for 18, 24, 16, 15, and 12.

SOLUTION:

Since no value occurs more than any other value, there is no mode.

A distribution can have one mode, more than one mode, or no mode. Also, the mean, median, and mode for a set of values most often differ somewhat.

PRACTICE

1. Find the mean, median, and mode for the number of sick days nine employees used last year. The data are 3, 6, 8, 2, 0, 5, 7, 8, and 5.
2. Find the mean, median, and mode for the number of rooms seven hotels in a large city have. The data are 332, 256, 300, 275, 216, 314, and 192.
3. Find the mean, median, and mode for the number of tornadoes that occurred in a specific state over the last 5 years. The data are 18, 6, 3, 9, and 10.
4. Find the mean, median, and mode for the number of items 9 people purchased at the express checkout register. The data are 12, 8, 6, 1, 5, 4, 6, 2, and 6.
5. Find the mean, median, and mode for the ages of 10 children who participated in a field trip to the zoo. The ages are 7, 12, 11, 11, 5, 8, 11, 7, 8, and 6.

ANSWERS

1. $\text{Mean} = \dfrac{3+6+8+2+0+5+7+8+5}{0} = \dfrac{44}{9} = 4.89$

   Median = 5
   Mode = 5 and 8

2. $\text{Mean} = \dfrac{332+256+300+275+216+314+192}{7} = \dfrac{1885}{7} = 269.29$

   Median = 275
   Mode = None

3. $\text{Mean} = \dfrac{18+6+3+9+10}{5} = \dfrac{46}{5} = 9.2$

   Median = 9
   Mode = None

4. $\text{Mean} = \dfrac{12+8+6+1+5+4+6+2+6}{9} = \dfrac{50}{9} = 5.56$

   Median = 6
   Mode = 6

5. $\text{Mean} = \dfrac{7+12+11+11+5+8+11+7+8+6}{10} = \dfrac{86}{10} = 8.6$

   Median = 8
   Mode = 11

Measures of Variability

In addition to measures of average, statisticians are interested in measures of variation. One measure of variability is called the *range*. The **range** is the difference between the largest data value and the smallest data value.

EXAMPLE: Find the range for 27, 32, 18, 16, 19, and 40.

SOLUTION:

Since the largest data value is 40 and the smallest data value is 16, the range is $40 - 16 = 24$.

Another measure that is also used as a measure of variability for individual data values is called the standard deviation. This measure was also used in Chapter 7.

The steps for computing the standard deviation for individual data values are

Step 1: Find the mean.
Step 2: Subtract the mean from each value and square the differences.
Step 3: Find the sum of the squares.
Step 4: Divide the sum by the number of data values minus one.
Step 5: Take the square root of the answer.

EXAMPLE: Find the standard deviation for 32, 18, 15, 24, and 11.

SOLUTION:

Step 1: Find the mean: $\dfrac{32 + 18 + 15 + 24 + 11}{5} = \dfrac{100}{5} = 20$

Step 2: Subtract the mean from each value and square the differences:

$$
\begin{array}{ll}
32 - 20 = 12 & 12^2 = 144 \\
18 - 20 = -2 & (-2)^2 = 4 \\
15 - 20 = -5 & (-5)^2 = 25 \\
24 - 20 = 4 & (4)^2 = 16 \\
11 - 20 = -9 & (-9)^2 = 81
\end{array}
$$

Step 3: Find the sum of the squares:
$144 + 4 + 25 + 16 + 81 = 270$
Step 4: Divide 270 by $5 - 1$ or 4: $270 \div 4 = 67.5$
Step 5: Take the square root of the answer $\sqrt{67.5} = 8.22$ (rounded)

The standard deviation is 8.22.

Recall from Chapter 7 that most data values fall within 2 standard deviations of the mean. In this case, $20 \pm 2 \cdot (8.22)$ is $3.56 <$ most values < 36.44. Looking at the data, you can see all the data values fall between 3.56 and 36.44.

EXAMPLE: Find the standard deviation for the number of minutes 10 people waited in a checkout line at a local supermarket. The times in minutes are 3, 0, 8, 2, 5, 6, 1, 4, 1, and 0.

SOLUTION:

Step 1: Find the mean: $\frac{3+0+8+2+5+6+1+4+1+0}{10} = \frac{30}{10} = 3$

Step 2: Subtract and square:

$$
\begin{array}{lr}
3-3=0 & 0^2=0 \\
0-3=-3 & (-3)^2=9 \\
8-3=5 & 5^2=25 \\
2-3=-1 & (-1)^2=1 \\
5-3=2 & 2^2=4 \\
6-3=3 & 3^2=9 \\
1-3=-2 & (-2)^2=4 \\
4-3=1 & 1^2=1 \\
1-3=-2 & (-2)^2=4 \\
0-3=-3 & (-3)^2=9
\end{array}
$$

Step 3: Find the sum: $0+9+25+1+4+9+4+1+4+9=66$
Step 4: Divide by 9: $66 \div 9 = 7.33$
Step 5: Take the square root: $\sqrt{7.33} = 2.71$ (rounded)

The standard deviation is 2.71.

PRACTICE

1. Twelve students were given a history test and the times (in minutes) they took to complete the test are shown: 8, 12, 15,16, 14, 10, 10, 11, 13, 15, 9, 11. Find the range and standard deviation.

2. Eight students were asked how many hours it took them to write a research paper. Their times (in hours) are shown: 6, 10, 3, 5, 7, 8, 2, 7. Find the range and standard deviation.
3. The high temperatures for 10 selected cities are shown: 32, 19, 57, 48, 44, 50, 42, 49, 53, 46. Find the range and standard deviation.
4. The times in minutes it took a driver to get to work last week are shown: 32, 35, 29, 31, 33. Find the range and standard deviation.
5. The number of hours 8 part-time employees worked last week is shown: 26, 28, 15, 25, 32, 36, 19, 11. Find the range and standard deviation.

ANSWERS

1. The range is $16 - 8 = 8$.

   The mean is $\frac{8+12+15+16+14+10+10+11+13+15+9+11}{12} = \frac{144}{12} = 12.$

   The standard deviation is

$$\begin{array}{rr} 8-12=-4 & (-4)^2=16 \\ 12-12=0 & 0^2=0 \\ 15-12=3 & 3^2=9 \\ 16-12=4 & 4^2=16 \\ 14-12=2 & 2^2=4 \\ 10-12=-2 & (-2)^2=4 \\ 10-12=-2 & (-2)^2=4 \\ 11-12=-1 & (-1)^2=1 \\ 13-12=1 & 1^2=1 \\ 15-12=3 & 3^2=9 \\ 9-12=-3 & (-3)^2=9 \\ 11-12=-1 & (-1)^2=1 \\ \hline & 74 \end{array}$$

$$\frac{74}{11} = 6.73 \qquad \sqrt{6.73} = 2.59 \text{ (rounded)}$$

2. Range $= 10 - 2 = 8$

$$\text{Mean} = \frac{6 + 10 + 3 + 5 + 7 + 8 + 2 + 7}{8} = \frac{48}{8} = 6$$

$$\begin{array}{lr}
6 - 6 = 0 & 0^2 = 0 \\
10 - 6 = 4 & 4^2 = 16 \\
3 - 6 = -3 & (-3)^2 = 9 \\
5 - 6 = -1 & (-1)^2 = 1 \\
7 - 6 = 1 & 1^2 = 1 \\
8 - 6 = 2 & 2^2 = 4 \\
2 - 6 = -4 & (-4)^2 = 16 \\
7 - 6 = 1 & 1^2 = \underline{1} \\
 & 48
\end{array}$$

$$\frac{48}{7} = 6.86 \quad \sqrt{6.86} = 2.62 \text{ (rounded)}$$

3. Range $= 57 - 19 = 38$

$$\text{Mean} = \frac{32 + 19 + 57 + 48 + 44 + 50 + 42 + 49 + 53 + 46}{10} =$$

$$\frac{440}{10} = 44$$

$$\begin{array}{lr}
32 - 44 = -12 & (-12)^2 = 144 \\
19 - 44 = -25 & (-25)^2 = 625 \\
57 - 44 = 13 & 13^2 = 169 \\
48 - 44 = 4 & 4^2 = 16 \\
44 - 44 = 0 & 0^2 = 0 \\
50 - 44 = 6 & 6^2 = 36 \\
42 - 44 = -2 & (-2)^2 = 4 \\
49 - 44 = 5 & 5^2 = 25 \\
53 - 44 = 9 & 9^2 = 81 \\
46 - 44 = 2 & 2^2 = \underline{4} \\
 & 1104
\end{array}$$

$$\frac{1104}{9} = 122.67 \qquad \sqrt{122.67} = 11.08$$

4. Range $= 35 - 29 = 6$

$$\text{Mean} = \frac{32 + 35 + 29 + 31 + 33}{5} = \frac{160}{5} = 32$$

$32 - 32 = 0$	$0^2 = 0$
$35 - 32 = 3$	$3^2 = 9$
$29 - 32 = -3$	$(-3)^2 = 9$
$31 - 32 = -1$	$(-1)^2 = 1$
$33 - 32 = 1$	$1^2 = 1$
	20

$\frac{20}{4} = 5 \qquad \sqrt{5} = 2.24$ (rounded)

5. Range $= 36 - 11 = 25$

$$\text{Mean} = \frac{26 + 28 + 15 + 25 + 32 + 36 + 19 + 11}{8} = \frac{192}{8} = 24$$

$26 - 24 = 2$	$2^2 = 4$
$28 - 24 = 4$	$4^2 = 16$
$15 - 24 = -9$	$(-9)^2 = 81$
$25 - 24 = 1$	$1^2 = 1$
$32 - 24 = 8$	$8^2 = 64$
$36 - 24 = 12$	$12^2 = 144$
$19 - 24 = -5$	$(-5)^2 = 25$
$11 - 24 = -13$	$(-13)^2 = 169$
	504

$\frac{504}{7} = 72 \qquad \sqrt{72} = 8.49$ (rounded)

The Normal Distribution

Recall from Chapter 7 that a continuous random variable can assume all values between any two given values. For example, the heights of adult males is a continuous random variable since a person's height can be any number. We are, however, limited by our measuring instruments. The variable temperature is a continuous variable since temperature can assume any numerical value between any two given numbers. Many continuous variables can be represented by formulas and graphs or curves. These curves represent probability distributions. In order to find probabilities for values of a variable, the area under the curve between two given values is used.

One of the most often used continuous probability distributions is called the **normal probability distribution**. Many variables are approximately normally distributed and can be represented by the normal distribution. It is important to realize that the normal distribution is a perfect theoretical mathematical curve but no real-life variable is perfectly normally distributed.

The real-life normally distributed variables can be described by the theoretical normal distribution. This is not so unusual when you think about it. Consider the wheel. It can be represented by the mathematically perfect circle, but no real-life wheel is perfectly round. The mathematics of the circle, then, is used to describe the wheel.

The normal distribution has the following properties:

1. It is bell-shaped.
2. The mean, median, and mode are at the center of the distribution.
3. It is symmetric about the mean. (This means that it is a reflection of itself if a mean was placed at the center.)
4. It is continuous; i.e., there are no gaps.
5. It never touches the x axis.
6. The total area under the curve is 1 or 100%.
7. About 0.68 or 68% of the area under the curve falls within one standard deviation on either side of the mean. (Recall that μ is the symbol for the mean and σ is the symbol for the standard deviation.)
   About 0.95 or 95% of the area under the curve falls within two standard deviations of the mean.
   About 1.00 or 100% of the area falls within three standard deviations of the mean. (Note: It is somewhat less than 100%, but for simplicity, 100% will be used here.) See Figure 9-1.

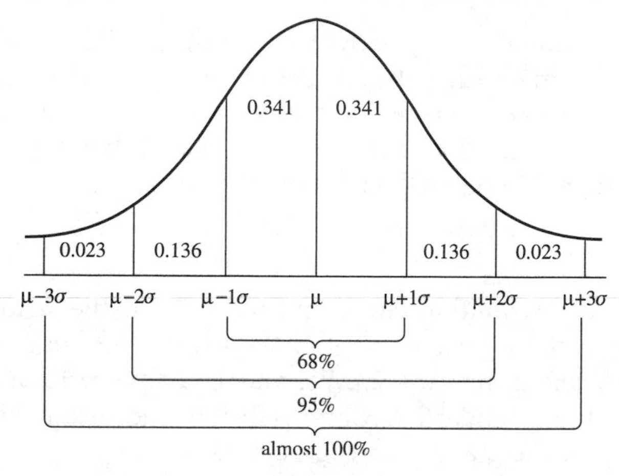

Fig. 9-1.

EXAMPLE: The mean commuting time between a person's home and office is 24 minutes. The standard deviation is 2 minutes. Assume the variable is normally distributed. Find the probability that it takes a person between 24 and 28 minutes to get to work.

SOLUTION:

Draw the normal distribution and place the mean, 24, at the center. Then place the mean plus one standard deviation (26) to the right, the mean plus two standard deviations (28) to the right, the mean plus three standard deviations (30) to the right, the mean minus one standard deviation (22) to the left, the mean minus two standard deviations (20) to the left, and the mean minus three standard deviations (18) to the left, as shown in Figure 9-2.

Using the areas shown in Figure 9-1, the area under the curve between 24 and 28 minutes is $0.341 + 0.136 = 0.477$ or 47.7%. Hence the probability that the commuter will take between 24 and 28 minutes is about 48%.

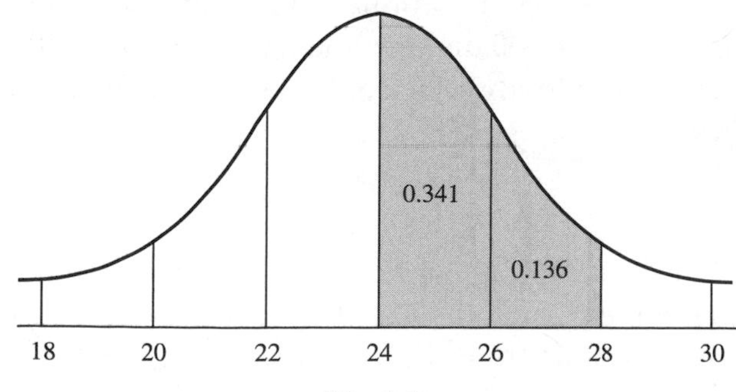

Fig. 9-2.

EXAMPLE: According to a study by A.C. Neilson, children between 2 and 5 years of age watch an average of 25 hours of television per week. Assume the variable is approximately normally distributed with a standard deviation of 2. If a child is selected at random, find the probability that the child watched more than 27 hours of television per week.

SOLUTION:

Draw the normal distribution curve and place 25 at the center; then place 27, 29, and 31 to the right corresponding to one, two, and three standard deviations above the mean, and 23, 21, and 19 to the left corresponding to one, two, and three standard deviations below the mean. Now place the areas (percents) on the graph. See Figure 9-3.

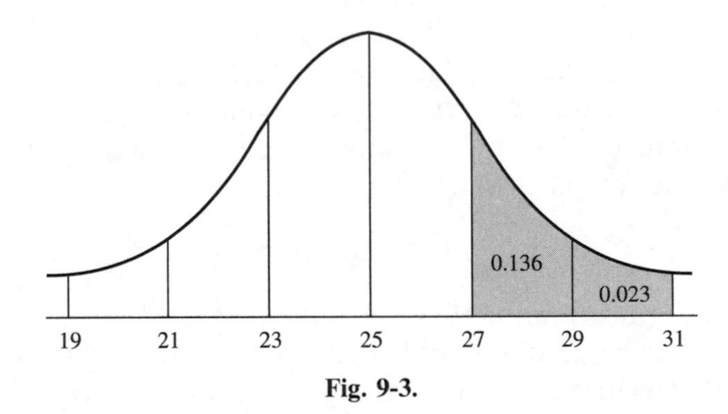

Fig. 9-3.

Since we are finding the probabilities for the number of hours greater than 27, add the areas of $0.136 + 0.023 = 0.159$ or 15.9%. Hence, the probability is about 16%.

EXAMPLE: The scores on a national achievement exam are normally distributed with a mean of 500 and a standard deviation of 100. If a student who took the exam is randomly selected, find the probability that the student scored below 600.

SOLUTION:

Draw the normal distribution curve and place 500 at the center. Place 600, 700, and 800 to the right and 400, 300, and 200 to the left, corresponding to one, two, and three standard deviations above and below the mean respectively. Fill in the corresponding areas. See Figure 9-4.

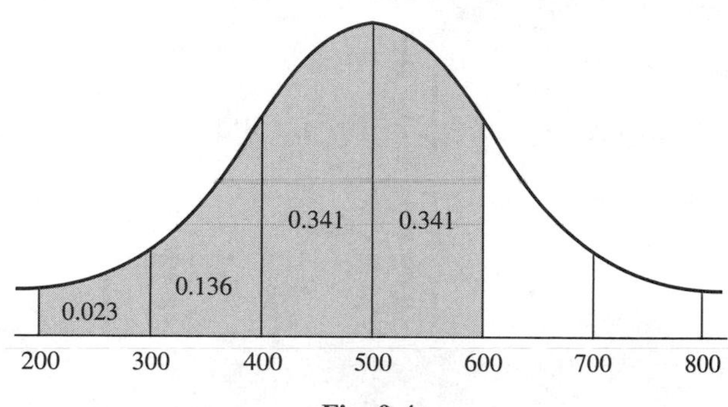

Fig. 9-4.

Since we are interested in the probability of a student scoring less than 600, add $0.341 + 0.341 + 0.136 + 0.023 = 0.841 = 84.1\%$. Hence, the probability of a randomly selected student scoring below 600 is 84%.

PRACTICE

1. To qualify to attend a fire academy, an applicant must take a written exam. If the mean of all test scores is 80 and the standard deviation is 5, find the probability that a randomly selected applicant scores between 75 and 95. Assume the test scores are normally distributed.
2. The average time it takes an emergency service to respond to calls in a certain municipality is 13 minutes. If the standard deviation is 3 minutes, find the probability that for a randomly selected call, the service takes less than 10 minutes. Assume the times are normally distributed.
3. If the measure of systolic blood pressure is normally distributed with a mean of 120 and a standard deviation of 10, find the probability that a randomly selected person will have a systolic blood pressure below 140. Assume systolic blood pressure is normally distributed.
4. If an automobile gets an average of 25 miles per gallon on a trip and the standard deviation is 2 miles per gallon, find the probability that on a randomly selected trip, the automobile will get between 21 and 29 miles per gallon. Assume the variable is normally distributed.
5. If adult Americans spent on average \$60 per year for books and the standard deviation of the variable is \$5, find the probability that a randomly selected adult spent between \$50 and \$65 last year on books. Assume the variable is normally distributed.

ANSWERS

1. The required area is shown in Figure 9-5.

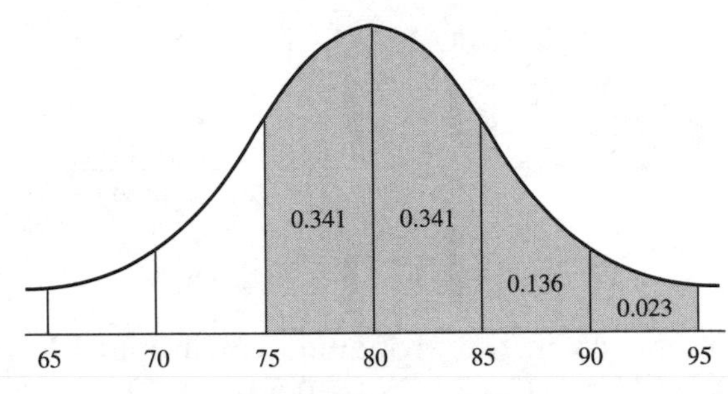

Fig. 9-5.

Probability = 0.341 + 0.341 + 0.136 + 0.023 = 0.841 or 84.1%

2. The required area is shown in Figure 9-6.

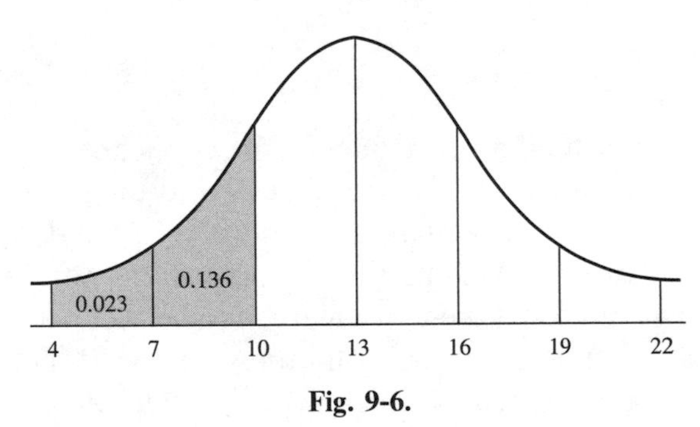

Fig. 9-6.

Probability = 0.023 + 0.136 = 0.159 or 15.9%

3. The required area is shown in Figure 9-7.

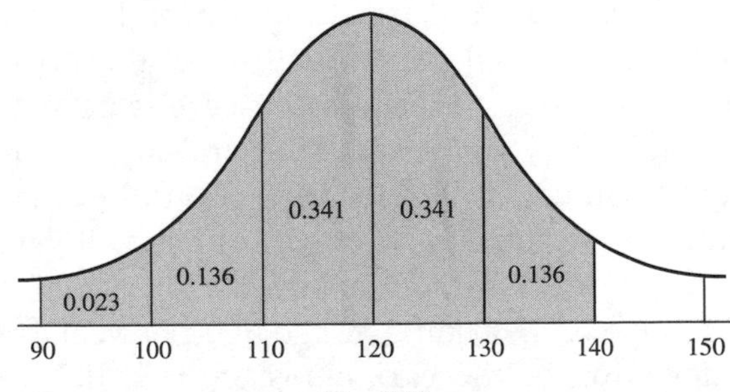

Fig. 9-7.

Probability = 0.023 + 0.136 + 0.341 + 0.341 + 0.136 = 0.977 or 97.7%

4. The required area is shown in Figure 9-8.

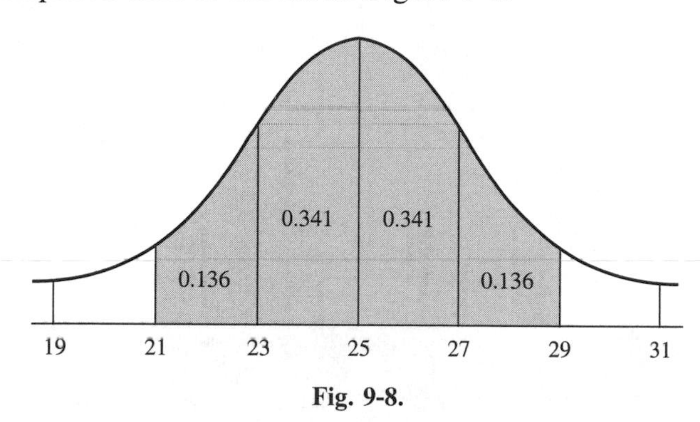

Fig. 9-8.

Probability = 0.136 + 0.341 + 0.341 + 0.136 = 0.954 or 95.4%

5. The required area is shown in Figure 9-9.

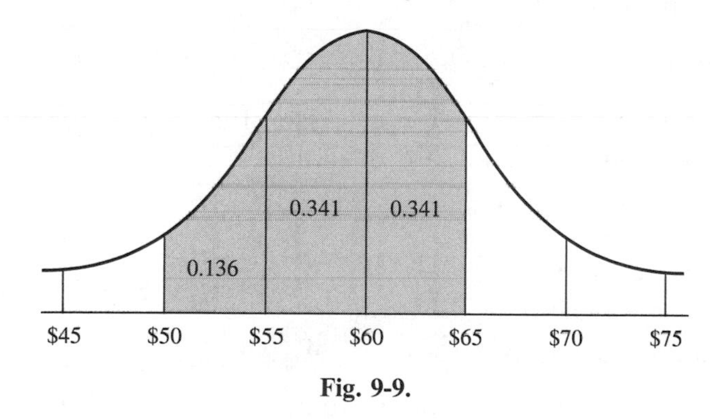

Fig. 9-9.

Probability = 0.136 + 0.341 + 0.341 = 0.818 or 81.8%

The Standard Normal Distribution

The normal distribution can be used as a model to solve many problems about variables that are approximately normally distributed. Since each variable has its own mean and standard deviation, statisticians use what is called the *standard normal distribution* to solve the problems.

The **standard normal distribution** has all the properties of a normal distribution, but the mean is zero and the standard deviation is one. See Figure 9-10.

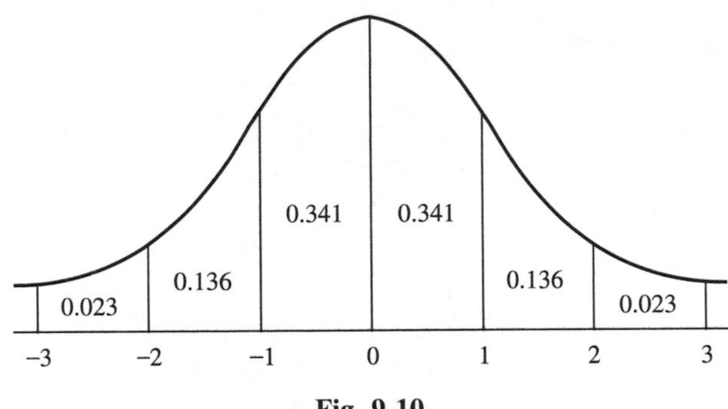

Fig. 9-10.

A value for any variable that is approximately normally distributed can be transformed into a standard normal value by using the following formula:

$$z = \frac{\text{value} - \text{mean}}{\text{standard deviation}}$$

The standard normal values are called ***z values*** or ***z* scores**.

EXAMPLE: Find the corresponding z value for a value of 18 if the mean of a variable is 12 and the standard deviation is 4.

SOLUTION:

$$z = \frac{\text{value} - \text{mean}}{\text{standard deviation}} = \frac{18 - 12}{4} = \frac{6}{4} = 1.5$$

Hence the z value of 1.5 corresponds to a value of 18 for an approximately normal distribution which has a mean of 12 and a standard deviation of 4. z values are negative for values of variables that are below the mean.

EXAMPLE: Find the corresponding z value for a value of 9 if the mean of a variable is 12 and the standard deviation is 4.

SOLUTION:

$$z = \frac{\text{value} - \text{mean}}{\text{standard deviation}} = \frac{9 - 12}{4} = -\frac{3}{4} = -0.75$$

Hence in this case a value of 9 is equivalent to a z value of −0.75.

In addition to finding probabilities for values that are between zero, one, two, and three standard deviations of the mean, probabilities for other values

can be found by converting them to z values and using the standard normal distribution.

Areas between any two given z values under the standard normal distribution curve can be found by using calculus instead; however, tables for specific z values can be found in any statistics textbook. An abbreviated table of areas is shown in Table 9-1.

Table 9-1 Approximate Cumulative Areas for the Standard Normal Distribution

z	**Area**	z	**Area**	z	**Area**	z	**Area**
−3.0	.001	−1.5	.067	0.0	.500	1.5	.933
−2.9	.002	−1.4	.081	0.1	.540	1.6	.945
−2.8	.003	−1.3	.097	0.2	.579	1.7	.955
−2.7	.004	−1.2	.115	0.3	.618	1.8	.964
−2.6	.005	−1.1	.136	0.4	.655	1.9	.971
−2.5	.006	−1.0	.159	0.5	.692	2.0	.977
−2.4	.008	−0.9	.184	0.6	.726	2.1	.982
−2.3	.011	−.08	.212	0.7	.758	2.2	.986
−2.2	.014	−0.7	.242	0.8	.788	2.3	.989
−2.1	.018	−0.6	.274	0.9	.816	2.4	.992
−2.0	.023	−0.5	.309	1.0	.841	2.5	.994
−1.9	.029	−0.4	.345	1.1	.864	2.6	.995
−1.8	.036	−0.3	.382	1.2	.885	2.7	.997
−1.7	.045	−0.2	.421	1.3	.903	2.8	.997
−1.6	.055	−0.1	.460	1.4	.919	2.9	.998
						3.0	.999

This table gives the approximate cumulative areas for z values between -3 and $+3$. The next three examples will show how to find the area (and corresponding probability in decimal form).

EXAMPLE: Find the area under the standard normal distribution curve to the left of $z = 1.3$.

SOLUTION:

The area is shown in Figure 9-11.

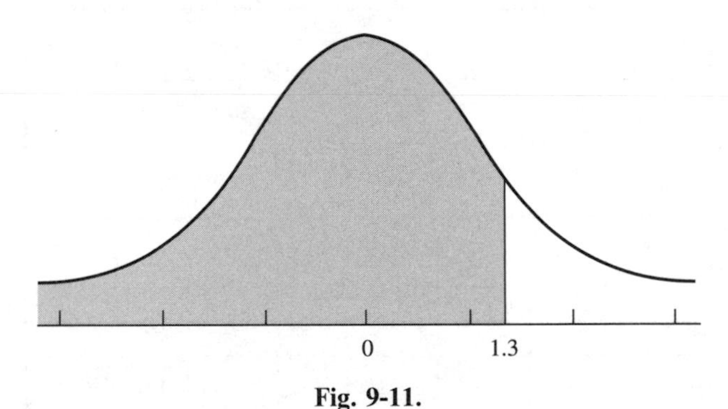

Fig. 9-11.

In order to find the area under the standard normal distribution curve to the left of any given z value, just look it up directly in Table 9-1. The area is 0.903 or 90.3%.

EXAMPLE: Find the area under the standard normal distribution curve between $z = -1.6$ and $z = 0.8$.

SOLUTION:

The area is shown in Figure 9-12.

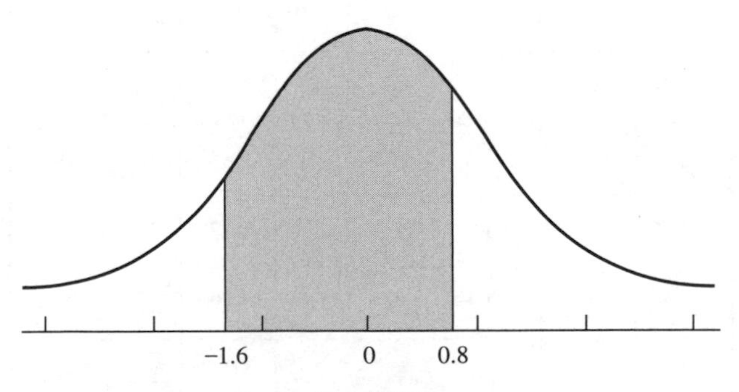

Fig. 9-12.

To find the area under the standard normal distribution curve between any given two z values, look up the areas in Table 9-1 and subtract the smaller area from the larger. In this case the area corresponding to $z = -1.6$ is 0.055, and the area corresponding to $z = 0.8$ is 0.788, so the area between $z = -1.6$ and $z = 0.8$ is $0.788 - 0.055 = 0.733 = 73.3\%$. In other words, 73.3% of the area under the standard normal distribution curve is between $z = -1.6$ and $z = 0.8$.

EXAMPLE: Find the area under the standard normal distribution curve to the right of $z = -0.5$.

SOLUTION:

The area is shown in Figure 9-13.

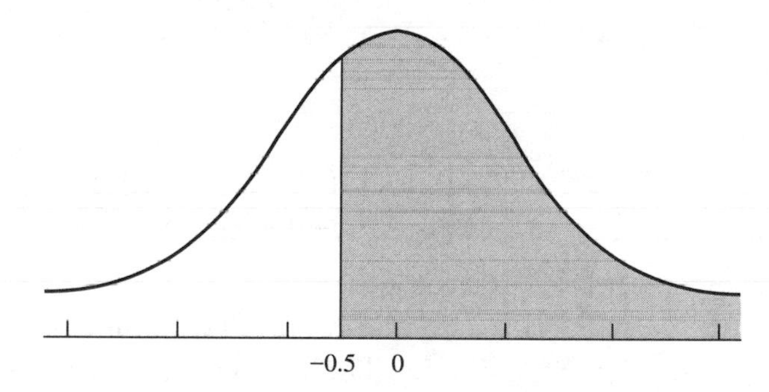

Fig. 9-13.

To find the area under the standard normal distribution curve to the right of any given z value, look up the area in the table and subtract that from 1. The area corresponding to $z = 0.5$ is 0.309. Hence $1 - 0.309 = 0.691$. The area to the right of $z = 0.5$ is 0.691. In other words, 69.1% of the area under the standard normal distribution curve lies to the right of $z = -0.5$.

Using Table 9-1 and the formula for transforming values for variables that are approximately normally distributed, you can find the probabilities of various events.

EXAMPLE: The scores on a national achievement exam are normally distributed with a mean of 500 and a standard deviation of 100. If a student is selected at random, find the probability that the student scored below 680.

SOLUTION:

Find the z value for 680:

$$z = \frac{\text{value} - \text{mean}}{\text{standard deviation}} = \frac{680 - 500}{100} = \frac{180}{100} = 1.8$$

Draw a figure and shade the area. See Figure 9-14.

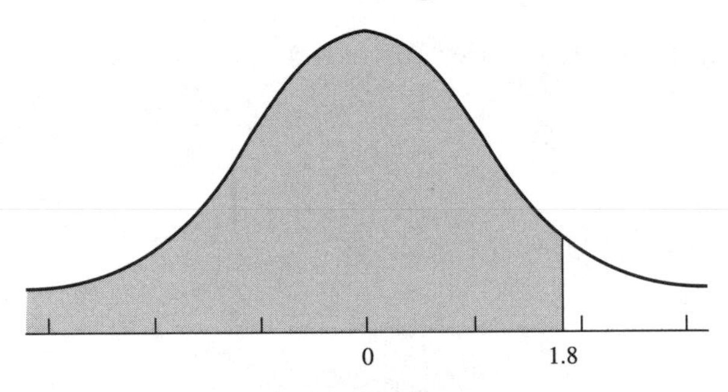

Fig. 9-14.

Look up 1.8 in Table 9-1 and find the area. It is 0.964. Hence the probability that a randomly selected student scores below 680 is 0.964 or 96.4%.

EXAMPLE: The average life of a certain brand of automobile tires is 24,000 miles under normal driving conditions. The standard deviation is 2000 miles, and the variable is approximately normally distributed. For a randomly selected tire, find the probability that it will last between 21,800 miles and 25,400 miles.

SOLUTION:

Find the two z values using the formula $z = \dfrac{\text{value} - \text{mean}}{\text{standard deviation}}$.

The z value for 21,800 miles is

$$z = \frac{21{,}800 - 24{,}000}{2000} = \frac{-2200}{2000} = -1.1$$

The corresponding area from Table 9-1 for -1.1 is 0.136.

The z value for 25,400 miles is

$$z = \frac{25{,}400 - 24{,}000}{2000} = 0.7$$

The corresponding area from Table 9-1 for 0.7 is 0.758.

Draw and label the normal distribution curve. Shade in the area. See Figure 9-15.

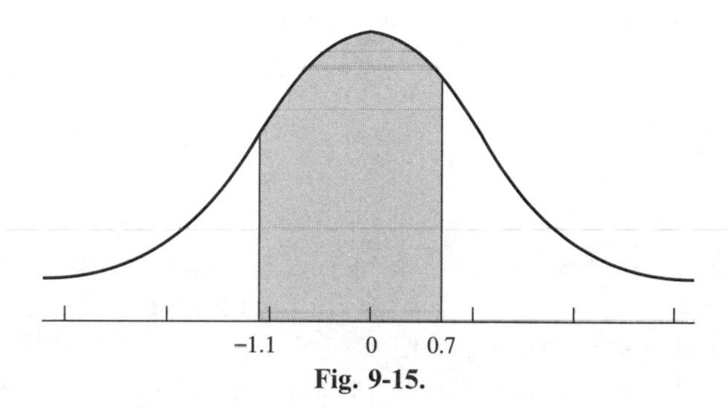

Fig. 9-15.

Subtract the smaller area from the larger area, $0.758 - 0.136 = 0.622$. Hence the probability that a randomly selected tire will last between 21,800 and 25,400 miles is 0.622 or 62.2%.

EXAMPLE: The average time it takes college freshmen to complete a reasoning skills test is 24 minutes. The standard deviation is 5 minutes. If a randomly selected freshman takes the exam, find the probability that he or she takes more than 32 minutes to complete the test. Assume the variable is normally distributed.

SOLUTION:

Find the z value for 32 minutes:

$$z = \frac{\text{value} - \text{mean}}{\text{standard deviation}} = \frac{32 - 24}{5} = \frac{8}{5} = 1.6$$

Draw the standard normal distribution and label it as shown. Shade the appropriate area. See Figure 9-16.

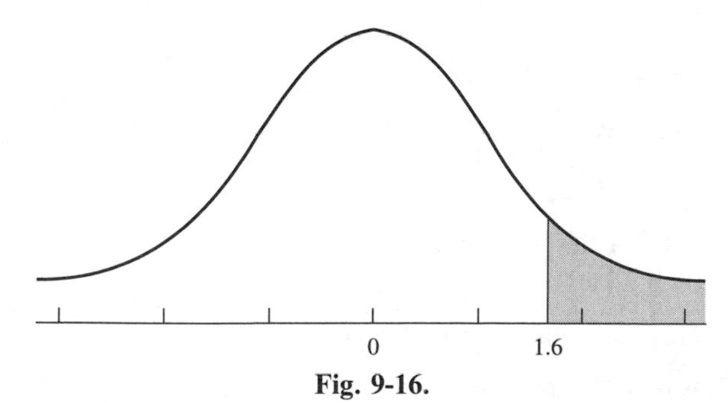

Fig. 9-16.

The area for $z = 1.6$ from Table 9-1 is 0.945. Subtract the area from 1. $1.00 - 0.945 = 0.055$. Hence the probability that it will take a randomly selected student longer than 32 minutes to complete the test is 0.055 or 5.5%.

PRACTICE

1. In order to qualify for a position, an applicant must score 86 or above on a skills test. If the test scores are normally distributed with a mean of 80 and a standard deviation of 4, find the probability that a randomly selected applicant will qualify for the position.
2. If a brisk walk at 4 miles per hour burns an average of 300 calories per hour, find the probability that a person will burn between 260 and 290 calories if the person walks briskly for one hour. Assume the standard deviation is 20 and the variable is approximately normally distributed.
3. The average count for snow per year that a city receives is 40 inches. The standard deviation is 10 inches. Find the probability that next year the city will get less than 53 inches. Assume the variable is normally distributed.
4. If the average systolic blood pressure is 120 and the standard deviation is 10, find the probability that a randomly selected person will have a blood pressure less than 108. Assume the variable is normally distributed.
5. A survey found that on average adults watch 2.5 hours of television per day. The standard deviation is 0.5 hours. Find the probability that a randomly selected adult will watch between 2.2 and 2.8 hours per day. Assume the variable is normally distributed.

ANSWERS

1. $z = \frac{86 - 80}{4} = \frac{6}{4} = 1.5$

   The required area is shown in Figure 9-17.

   The area for $z = 1.5$ is 0.933. Since we are looking for the area greater than $z = 1.5$, subtract the table value from 1: $1 - 0.933 = 0.067$. Hence the probability is 0.067 or 6.7%.

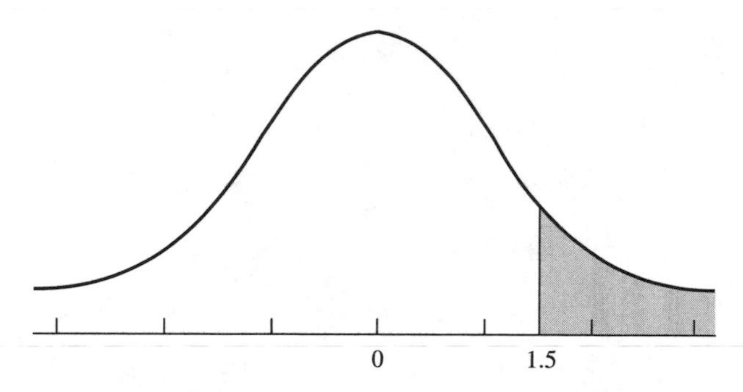

Fig. 9-17.

2. $z = \frac{290 - 300}{20} = -\frac{10}{20} = -0.5$

$z = \frac{260 - 300}{20} = -\frac{40}{20} = -2$

The required area is shown in Figure 9-18.

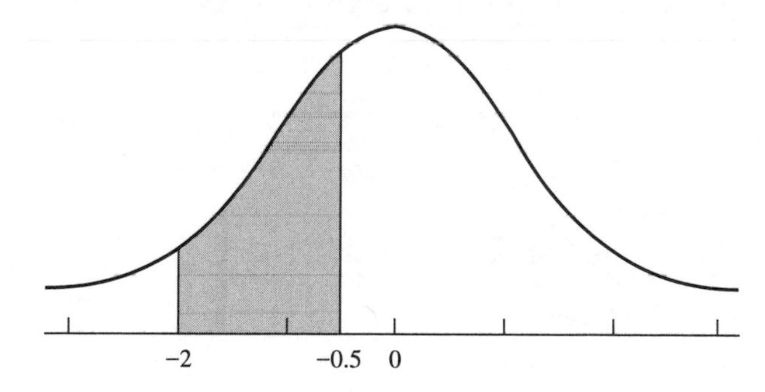

Fig. 9-18.

The area for −0.5 is 0.309. The area for −2 is 0.023. Since we are looking for the area between $z = -0.5$ and $z = -2$, subtract the areas. Hence the probability that a person burns between 260 and 290 calories is $0.309 - 0.023 = 0.286$ or 28.6%.

3. $z = \frac{53 - 40}{10} = \frac{13}{10} = 1.3$

The required area is shown in Figure 9-19.

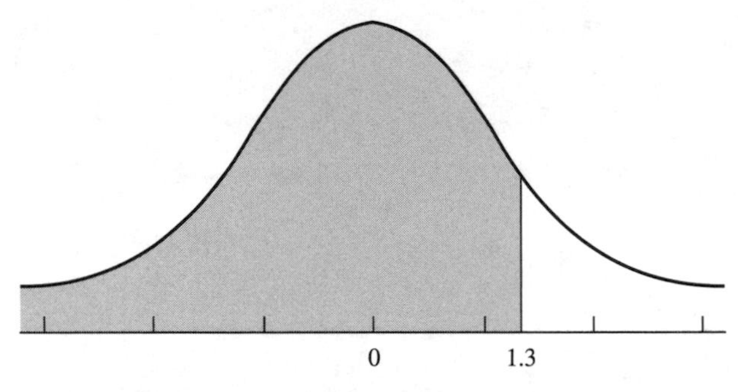

Fig. 9-19.

The area for $z = 1.3$ is 0.903. Since we are looking for an area less than $z = 1.3$, use the value found in the table. Hence the probability that the city receives less than 53 inches of snow is 0.903 or 90.3%.

4. $z = \dfrac{108 - 120}{10} = -\dfrac{12}{10} = -1.2$

The required area is shown in Figure 9-20.

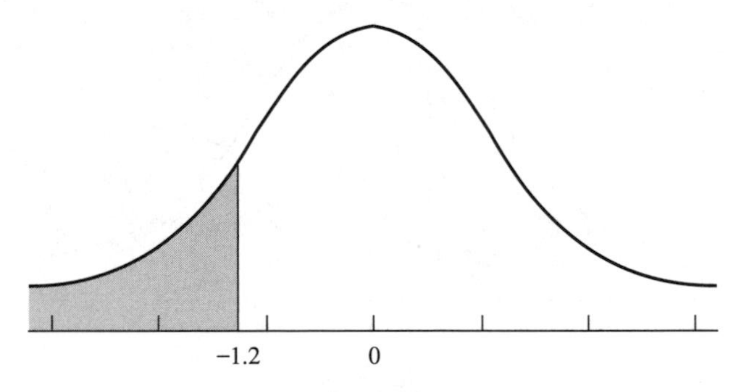

Fig. 9-20.

The area for $z = -1.2$ is 0.115. Since we are looking for the area less than $z = -1.2$, use the value in the table. Hence the probability that a person has a systolic blood pressure less than 108 is 0.115 or 11.5%.

5. $z = \dfrac{2.8 - 2.5}{0.5} = \dfrac{0.3}{0.5} = 0.6$

$z = \dfrac{2.2 - 2.5}{0.5} = -\dfrac{0.3}{0.5} = -0.6$

The required area is shown in Figure 9-21.

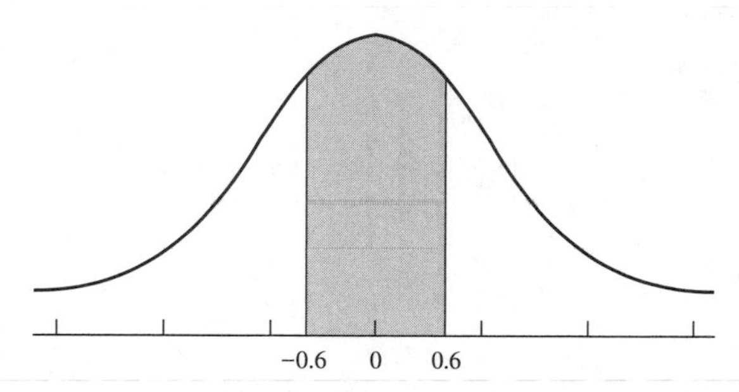

Fig. 9-21.

The area for $z = 0.6$ is 0.726. The area for $z = -0.6$ is 0.274. Since we are looking for the area between −0.6 and 0.6, subtract the areas: $0.726 - 0.274 = 0.452$ or 45.2%. Hence the probability that an adult will watch between 2.2 and 2.8 hours of television per day is 0.452 or 45.2%.

Summary

Statistics is a branch of mathematics that uses probability. Statistics uses data to analyze, summarize, make inferences, and draw conclusions from data. There are three commonly used measures of average. They are the mean, median, and mode. The mean is the sum of the data values divided by the number of data values. The median is the midpoint of the data values when they are arranged in numerical order. The mode is the data value that occurs most often.

There are two commonly used measures of variability. They are the range and standard deviation. The range is the difference between the smallest data value and the largest data value. The standard deviation is the square root of the average of the squares of the differences of each value from the mean.

Many variables are approximately normally distributed and the standard normal distribution can be used to find probabilities for various situations involving values of these variables.

The standard normal distribution is a continuous, bell-shaped curve such that the mean, median, and mode are at its center. It is also symmetrical about the mean. The mean is equal to zero and the standard deviation is equal to one. About 68% of the area under the standard normal distribution lies within one standard deviation of the mean, about 95% within two standard deviations, and about 100% within three standard deviations of the mean.

CHAPTER QUIZ

1. What is the mean of 18, 24, 6, 12, 15, and 12?

   a. 12
   b. 14.5
   c. 87
   d. 10.6

2. What is the median of 25, 15, 32, 43, 15, and 6?

   a. 20
   b. 15
   c. 22.8
   d. 23.5

3. What is the mode of 17, 19, 26, 43, 26, 15, and 14?

   a. 22.9
   b. 26
   c. 17
   d. None

4. What is the range of 32, 50, 14, 26, 18 and 25?

   a. 36
   b. 25
   c. 18
   d. 43

5. What is the standard deviation of 12, 18, 14, 22, and 21?

   a. 3.9
   b. 18.8
   c. 4.3
   d. 17.4

6. Which is **not** a property of the normal distribution?

   a. It is bell-shaped.
   b. It is continuous.
   c. The mean is at the center.
   d. It is not symmetrical about the mean.

7. The area under the standard normal distribution is

   a. 1 or 100%
   b. 0.5 or 50%
   c. Unknown
   d. Infinite

8. For the standard normal distribution,

   a. The mean = 1 and the standard deviation = 0.
   b. The mean = 1 and the standard deviation = 2.
   c. The mean = 0 and the standard deviation = 1.
   d. The mean and standard deviation can both vary.

9. What percent of the area under the standard normal distribution lies within two standard deviations of the mean?

   a. 68%
   b. 95%
   c. 100%
   d. Unknown

10. When the value of a variable is transformed into a standard normal variable, the new value is called a(n)

    a. x value
    b. y value
    c. 0 value
    d. z value

Use Table 9-1 to answer questions 11 through 15.

11. The area under the standard normal distribution to the right of $z = 0.9$ is

    a. 90.1%
    b. 18.4%
    c. 81.6%
    d. 10.2 %

12. The area under the standard normal distribution to the left of $z = -1.2$ is

    a. 88.5%
    b. 62.3%
    c. 48.7%
    d. 11.5%

13. The area under the standard normal distribution between $z=-1.7$ and $z=0.5$ is

    a. 69.2%
    b. 4.5%
    c. 35.6%
    d. 64.7%

14. An exam which is approximately normally distributed has a mean of 200 and a standard deviation of 20. If a person who took the exam is selected at random, find the probability that the person scored above 230.

    a. 93.3%
    b. 70.2%
    c. 6.7%
    d. 30.8%

15. The average height for adult females is 64 inches. Assume the variable is normally distributed with a standard deviation of 2 inches. If a female is randomly selected, find the probability that her height is between 62 and 66.8 inches.

    a. 92%
    b. 76%
    c. 32%
    d. 16%

Probability Sidelight

A BRIEF HISTORY OF THE NORMAL DISTRIBUTION

The applications of the normal distribution are many and varied. It is used in astronomy, biology, business, education, medicine, engineering, psychology, and many other areas. The development of the concepts of the normal distribution is quite interesting.

It is believed that the first mathematician to discover some of the concepts associated with the normal distribution was the English mathematician, Abraham de Moivre (1667–1754). He was born in France, but moved to England because of the French government's restrictions on religion and civil liberties. He supported himself by becoming a private tutor in mathematics.

He studied the probability of tossing coins, rolling dice, and other forms of gambling. In 1716, he wrote a book on gambling entitled *The Doctrine of Chances*. In addition to his tutoring, wealthy patrons came to him to find out what the payoff amount should be for various gambling games. He made many contributions to mathematics. In probability, he tossed a large number of coins many times and recorded the number of heads that resulted on each trial. He found that approximately 68% of the results fell within a predictable distance (now called the standard deviation) on either side of the mean and that 95% of the results fell within two predictable distances on either side of the mean. In addition, he noticed that the shape of the distribution was bell-shaped, and he derived the equation for the normal curve in 1733, but his work in this area of mathematics went relatively unnoticed for a long period of time.

In 1781, a French mathematician, Pierre Simon Laplace (1749–1827) was studying the gender of infants attempting to prove that the number of males born was slightly more than the number of females born. (This fact has been verified today.) Laplace noticed that the distribution of male births was also bell-shaped and that the outcomes followed a particular pattern. Laplace also developed a formula for the normal distribution, and it is thought that he was unaware of de Moivre's earlier work.

About 30 years later in 1809, a German mathematician, Carl Friedrich Gauss (1777–1855) deduced that the errors in the measurements of the planets due to imperfections in the lenses in telescopes and the human eye were approximately bell-shaped. The theory was called Gauss' Law of Error. Gauss developed a complex measure of variation for the data and also an equation for the normal distribution curve. The curve is sometimes called the Gaussian distribution in his honor. In addition to mathematics, Gauss also made many contributions to astronomy.

During the 1800s at least seven different measures of variation were used to describe distributions. It wasn't until 1893 that the statistician Karl Pearson coined the term "standard deviation."

Around 1830, researchers began to notice that the normal distribution could be used to describe other phenomena. For example, in 1846 Adolphe Quetelet (1796–1874) began to measure the chest sizes of Scottish soldiers. He was trying to develop the concept of the "average man," and found that the normal distribution curve was applicable to these measurements. Incidentally, Quetelet also developed the concept of body mass index, which is still used today.

A German experimental psychologist, Hermann Ebbinghaus (1855–1913) found that the normal distribution was applicable to measures of intelligence and memorization in humans.

Sir Francis Galton in 1889 invented a device that he called a "Quincunx." This device dropped beads through a series of pegs into slots whose heights resulted in a bell-shaped graph. It wasn't until 1924 that Karl Pearson found that de Moivre had discovered the formula for the normal distribution curve long before Laplace or Gauss.

Simulation

Introduction

Instead of studying actual situations that sometimes might be too costly, too dangerous, or too time consuming, researchers create similar situations using random devices so that they are less expensive, less dangerous or less time consuming. For example, pilots use flight simulators to practice on before they actually fly a real plane. Many video games use the computer to simulate real life sports situations such as baseball, football, or hockey.

Simulation techniques date back to ancient times when the game of chess was invented to simulate warfare. Modern techniques date to the mid-1940s when two physicists, John von Neuman and Stanislaw Ulam developed simulation techniques to study the behavior of neutrons in the design of atomic reactors.

Mathematical simulation techniques use random number devices along with probability to create conditions similar to those found in real life.

Random devices are items such as dice, coins, and computers or calculators. These devices generate what are called **random numbers**. For example, when a fair die is rolled, it generates the numbers one through six randomly. This means that the outcomes occur by chance and each outcome has the same probability of occurring.

Computers have played an important role in simulation since they can generate random numbers, perform experiments, and tally the results much faster than humans can. In this chapter, the concepts of simulation will be explained by using dice or coins.

The Monte Carlo Method

The Monte Carlo Method of simulation uses random numbers. The steps are

Step 1: List all possible outcomes of the experiment.
Step 2: Determine the probability of each outcome.
Step 3: Set up a correspondence between the outcomes of the experiment and random numbers.
Step 4: Generate the random numbers (i.e., roll the dice, toss the coin, etc.)
Step 5: Repeat the experiment and tally the outcomes.
Step 6: Compute any statistics and state the conclusions.

If an experiment involves two outcomes and each has a probability of $\frac{1}{2}$, a coin can be tossed. A head would represent one outcome and a tail the other outcome. If a die is rolled, an even number could represent one outcome and an odd number could represent the other outcome. If an experiment involves five outcomes, each with a probability of $\frac{1}{5}$, a die can be rolled. The numbers one through five would represent the outcomes. If a six is rolled, it is ignored.

For experiments with more than six outcomes, other devices can be used. For example, there are dice for games that have 5 sides, 8 sides, 10 sides, etc. (Again, the best device to use is a random number generator such as a computer or calculator or even a table of random numbers.)

EXAMPLE: Simulate the genders of a family with four children.

SOLUTION:

Four coins can be tossed. A head represents a male and a tail represents a female. For example, the outcome HTHH represents 3 boys and one girl.

Perform the experiment 10 times to represent the genders of the children of 10 families. (Note: The probability of a male or a female birth is not exactly $\frac{1}{2}$; however, it is close enough for this situation.) The results are shown next.

Trial	Outcome	Number of boys
1	TTHT	1
2	TTTT	0
3	HHTT	2
4	THTT	1
5	TTHT	1
6	HHHH	4
7	HTHH	3
8	THHH	3
9	THHT	2
10	THTT	1

Results:

No. of boys	0	1	2	3	4
No. of families	1	4	2	2	1

In this case, there was one family with no boys and one family with four boys. Four families had one boy and three girls, and two families had two boys. The average is 1.8 boys per family of four.

More complicated problems can be simulated as shown next.

EXAMPLE: Suppose a prize is given under a bottle cap of a soda; however, only one in five bottle caps has the prize. Find the average number of bottles that would have to be purchased to win the prize. Use 20 trials.

SOLUTION:

A die can be rolled until a certain arbitrary number, say 3, appears. Since the probability of getting a winner is $\frac{1}{5}$, the number of rolls will be tallied. The experiment can be done 20 times. (In general, the more times the experiment is performed, the better the approximation will be.) In this case, if a six is rolled, it is not counted. The results are shown next.

Trial	Number of rolls until a 3 was obtained
1	1
2	6
3	5
4	4
5	11
6	5
7	1
8	3
9	7
10	2
11	4
12	2
13	4
14	1
15	6
16	9
17	1
18	5
19	7
20	11

Now, the average of the number of rolls is 4.75.

EXAMPLE: A box contains 3 one dollar bills, 2 five dollar bills, and 1 ten dollar bill. A person selects a bill at random. Find the expected value of the bill. Perform the experiment 20 times.

SOLUTION:

A die can be rolled. If a 1, 2, or 3 comes up, assume the person wins \$1. If a 4 or 5 comes up, assume the person wins \$5. If a 6 comes up, assume the person wins \$10.

Trial	Number	Amount
1	3	$1
2	6	$10
3	3	$1
4	6	$10
5	4	$5
6	1	$1
7	6	$10
8	4	$5
9	4	$5
10	3	$1
11	6	$10
12	1	$1
13	2	$1
14	5	$5
15	5	$5
16	3	$1
17	1	$1
18	2	$1
19	6	$10
20	3	$1

The average of the amount won is \$4.25.

The theoretical average or expected value can be found by using the formula shown in Chapter 5.

$E(X) = \frac{1}{2}(\$1) + \frac{1}{3}(\$5) + \frac{1}{6}(\$10) = \$3.83$. Actually, I did somewhat better than average.

The Monty Hall problem is a probability problem based on a game played on the television show "Let's Make A Deal," hosted by Monty Hall.

Here's how it works. You are a contestant on a game show, and you are to select one of three doors. A valuable prize is behind one door, and no prizes are behind the other two doors. After you choose a door, the game show host opens one of the two doors that you did not select. The game show host knows which door contains the prize and always opens a door with no prize behind it. Then the host asks you if you would like to keep the door you originally selected or switch to the other unopened door. The question is "Do you have a better chance of winning the valuable prize if you switch or does it make no difference?"

At first glance, it looks as if it does not matter whether or not you switch since there are two doors and only one has the prize behind it. So the probability of winning is $\frac{1}{2}$ whether or not you switch. This type of reasoning is incorrect since it is actually better if you switch doors! Here's why.

Assume you select door A. If the prize is behind door C, the host opens door B, so if you switch, you win. If the prize is behind door B, the host opens door C, so if you switch, you win. If the prize is behind door A and no matter what door the host opens, if you switch, you lose. So by always switching, you have a $\frac{2}{3}$ chance of winning and a $\frac{1}{3}$ chance of losing. If you don't switch, you will have only a $\frac{1}{3}$ chance of winning no matter what. You can apply the same reasoning if you select door B or door C. If you always switch, the probability of winning is $\frac{2}{3}$. If you don't switch, the probability of winning is $\frac{1}{3}$.

You can simulate the game by using three cards, say an ace and two kings. Consider the ace the prize. Turn your back and have a friend arrange the cards face down on a table. Then select a card. Have your friend turn over one of the other cards, not the ace, of course. Then switch cards and see whether or not you win. Keep track of the results for 10, 20, or 30 plays. Repeat the game but this time, don't switch, and keep track of how many times you win. Compare the results!

You can also play the game by visiting this website:

http://www.stat.sc.edu/~west/javahtm/LetsMakeaDeal.html

PRACTICE

Use simulation to estimate the answer.

1. A basketball player makes $\frac{2}{3}$ of his foul shots. If he has two shots, find the probability that he will make at least one basket.
2. In a certain prize give away, you must spell the word "BIG" to win. Sixty percent of the tickets have a B, 20% of the tickets have an I, and 20% of the tickets have a G. If a person buys 5 tickets, find the probability that the person will win a prize.
3. Two people shoot clay pigeons. Gail has an 80% accuracy rate, while Paul has a 50% accuracy rate. The first person who hits the target wins. If Paul always shoots first, find the probability that he wins.
4. A person has 5 neckties and randomly selects one tie each work day. In a given work week of 5 days, find the probability that the person will wear the same tie two or more days a week.
5. Toss three dice. Find the probability of getting exactly two numbers that are the same (doubles).

ANSWERS

1. You will roll 2 dice. A success is getting a 1, 2, 3, or 4, and a miss is getting a 5 or 6. Perform the experiment 50 times, tally the successes, and divide by 50 to get the probability.
2. You can roll a dice 5 times or 5 dice at one time. Count 1, 2, or 3 as a B, count 4 as an I, and count 5 as a G. If you get a six, roll the die over. Perform 50 experiments. Tally the wins (i.e., every time you spell BIG); then divide by 50 to get the probability.
3. Roll a die. For the first shooter, use 1, 2, and 3 as a hit and 4, 5, and 6 as a miss. For the second shooter, use 1, 2, 3, and 4 as a hit and 5 as a miss. Ignore any sixes.
4. Use the numbers 1, 2, 3, 4, and 5 on a die to represent the ties. Ignore any sixes. Roll the die five times for each week and count any time you get the same number twice as selecting the same tie twice.
5. Roll three dice and count the doubles.

Summary

Random numbers can be used to simulate many real life situations. The basic method of simulation is the Monte Carlo method. The purpose of simulation is to duplicate situations that are too dangerous, too costly, or too time consuming to study in real life. Most simulation techniques are done on a computer. Computers enable the person to generate random numbers, tally the results, and perform any necessary computation.

CHAPTER QUIZ

1. Two people who developed simulation techniques are
   a. Fermat and Pascal
   b. Laplace and DeMoivre
   c. Von Neuman and Ulam
   d. Plato and Aristotle
2. Mathematical simulation techniques use _____ numbers.
   a. Prime
   b. Odd
   c. Even
   d. Random
3. The simulation techniques explained in this chapter use the _____ method.
   a. Monte Carlo
   b. Casino
   c. Coin/Die
   d. Tally
4. A coin can be used as a simulation device when there are two outcomes and each outcome has a probability of
   a. $\frac{1}{4}$
   b. $\frac{1}{2}$
   c. $\frac{1}{3}$
   d. $\frac{1}{6}$

5. Which device will **not** generate random numbers?
   a. Computer
   b. Abacus
   c. Dice
   d. Calculator

Probability Sidelight

PROBABILITY IN OUR DAILY LIVES

People engage in all sorts of gambles, not just betting money at a casino or purchasing a lottery ticket. People also bet their lives by engaging in unhealthy activities such as smoking, drinking, using drugs, and exceeding the speed limit when driving. A lot of people don't seem to care about the risks involved in these activities, or they don't understand the concepts of probability.

Statisticians (called actuaries) who work for insurance companies can calculate the probabilities of dying from certain causes. For example, based on the population of the United States, the risks of dying from various causes are shown here.

Motor vehicle accident	1 in 7000
Shot with a gun	1 in 10,000
Crossing a street	1 in 60,000
Struck by lightning	1 in 3,000,000
Shark attack	1 in 300,000,000

The risks of dying from various diseases are shown here.

Heart attack	1 in 400
Cancer	1 in 600
Stroke	1 in 2000

As you can see, the probability of dying from a disease is much higher than the probability of dying from an accident.

Another thing that people tend to do is fear situations or events that have a relatively small chance of happening and overlook situations or events that have a higher chance of happening. For example, James Walsh in his book entitled *How Risk Affects Your Everyday Life* states that if a person is 20% overweight, the loss of life expectancy is 900 days (about 3 years) whereas the loss of life expectancy from exposure to radiation emitted by nuclear power plants is 0.02 days. So you can see it is much more unhealthy being 20% overweight than it is living close to a nuclear power plant. One of the reasons for this phenomenon is that the media tends to sensationalize certain news events such as floods, hurricanes, and tornadoes and downplays other less newsworthy events such as smoking, drinking, and being overweight.

In summary, then, when you make a decision or plan a course of action based on probability, get the facts from a reliable source, weigh the consequences of each choice of action, and then make your decision. Be sure to consider as many alternatives as you can.

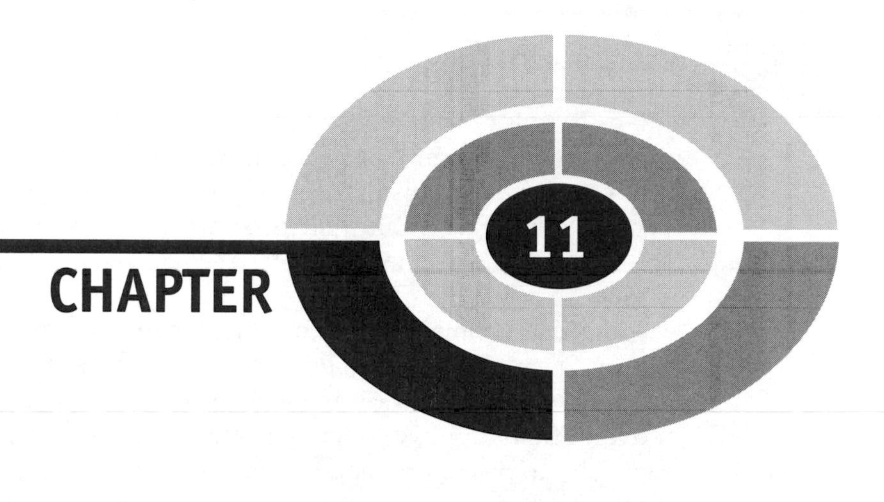

Game Theory

Introduction

Probability is used in what is called *game theory*. **Game theory** was developed by John von Neumann and is a mathematical analysis of games. In many cases, game theory uses probability. In a broad sense, game theory can be applied to sports such as football and baseball, video games, board games, gambling games, investment situations, and even warfare.

Two-Person Games

A simplified definition of a **game** is that it is a contest between two players that consists of rules on how to play and how to determine the winner. A game also consists of a *payoff*. A **payoff** is a reward for winning the game. In many cases it is money, but it could be points or even just the satisfaction of winning.

Most games consist of *strategies*. A **strategy** is a rule that determines a player's move or moves in order to win the game or maximize the player's payoff. When a game consists of the loser paying the winner, it is called a **zero sum game.** This means that the sum of the payoffs is zero. For example, if a person loses a game and that person pays the winner \$5, the loser's payoff is −\$5 and the winner's payoff is +\$5. Hence the sum of the payoff is −\$5 + \$5 = \$0.

Consider a simple game in which there are only two players and each player can make only a finite number of moves. Both players make a move simultaneously and the outcome or payoff is determined by the pair of moves. An example of such a game is called, "rock-paper-scissors." Here each player places one hand behind his or her back, and at a given signal, brings his or her hand out with either a fist, symbolizing "rock," two fingers out, symbolizing "scissors," or all five fingers out symbolizing "paper." In this game, scissors cut paper, so scissors win. A rock breaks scissors, so the rock wins, and paper covers rock, so paper wins. Rock–rock, scissors–scissors, and paper–paper are ties and neither person wins. Now suppose there are two players, say Player A and Player B, and they decide to play for \$1. The game can be symbolized by a rectangular array of numbers called a **payoff table**, where the rows represent Player A's moves and the columns represent Player B's moves. If Player A wins, he gets \$1 from Player B. If Player B wins, Player A pays him \$1, represented by −\$1. The payoff table for the game is

	Player B's Moves:		
Player A's Moves:	Rock	Paper	Scissors
Paper	0	−\$1	\$1
Rock	\$1	0	−\$1
Scissors	−\$1	\$1	0

This game can also be represented by a tree diagram, as shown in Figure 11-1.

Now consider a second game. Each player has two cards. One card is black on one side, and the other card is white on one side. The backs of all four cards are the same, so when a card is placed face down on a table, the color on the opposite side cannot be seen until it is turned over. Both players select a card and place it on the table face down; then they turn the cards over. If the result is two black cards, Player A wins \$5. If the result is two white cards, Player A wins \$1. If the results are one black card and

Player A's move | Player B's move | Payoff

Rock — Rock 0; Paper −$1; Scissors $1

Paper — Rock $1; Paper 0; Scissors −$1

Scissors — Rock −$1; Paper $1; Scissors 0

Fig. 11-1.

one white card, Player B wins $2.00. A payoff table for the game would look like this:

	Player B's Card:	
Player A's Card:	Black	White
Black	\$5	−\$2
White	−\$2	\$1

The tree diagram for the game is shown in Figure 11-2.

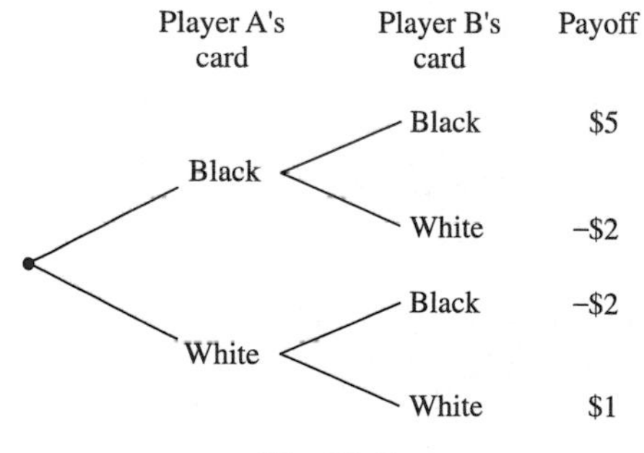

Fig. 11-2.

Player A thinks, "What about a strategy? I will play my black card and hope Player B plays her black card, and I will win \$5. But maybe Player B knows this and she will play her white card, and I will lose \$2. So, I better play my white card and hope Player B plays her white card, and I will win \$1. But she might realize this and play her black card! What should I do?"

In this case, Player A decides that he should play his black card some of the time and his white card some of the time. But how often should he play his black card?

This is where probability theory can be used to solve Player A's dilemma. Let $p =$ the probability of playing a black card on each turn; then $1 - p =$ the probability of playing a white card on each turn. If Player B plays her black card, Player A's expected profit is $\$5 \cdot p - \$2(1 - p)$. If Player B plays her white card, Player A's expected profit is $-\$2p + \$1(1 - p)$, as shown in the table.

	Player B's Card	
Player A's Card	Black	White
Black	$\$5p$	$-\$2p$
White	$-\$2(1-p)$	$\$1(1-p)$
	$\$5p - \$2(1-p)$	$-\$2p + \$1(1-p)$

Now in order to plan a strategy so that Player B cannot outthink Player A, the two expressions should be equal. Hence,

$$5p - 2(1 - p) = -2p + 1(1 - p)$$

Using algebra, we can solve for p:

$$5p - 2(1 - p) = -2p + 1(1 - p)$$

$$5p - 2 + 2p = -2p + 1 - p$$

$$7p - 2 = -3p + 1$$

$$7p + 3p - 2 = -3p + 3p + 1$$

$$10p - 2 = 1$$

$$10p - 2 + 2 = 1 + 2$$

$$10p = 3$$

$$\frac{10p}{10} = \frac{3}{10}$$

$$p = \frac{3}{10}$$

Hence, Player A should play his black card $\frac{3}{10}$ of the time and his white card $\frac{7}{10}$ of the time. His expected gain, no matter what Player B does, when $p=\frac{3}{10}$ is

$$5p - 2(1-p) = 5 \cdot \frac{3}{10} - 2\left(1 - \frac{3}{10}\right)$$
$$= \frac{15}{10} - 2\left(\frac{7}{10}\right)$$
$$= \frac{15}{10} - \frac{14}{10}$$
$$= \frac{1}{10} \text{ or } \$0.10$$

On average, Player A will win $0.10 per game no matter what Player B does.

Now Player B decides she better figure her expected loss no matter what Player A does. Using similar reasoning, the table will look like this when the probability that Player B plays her black card is s, and her white card with probability $1-s$.

	Player B's Card:		
Player's A Card:	Black	White	
White	$\$5s$	$-\$2(1-s)$	$\$5s - \$2(1-s)$
Black	$-\$2s$	$\$1(1-s)$	$-\$2(s) + \$1(1-s)$

Solving for s when both expressions are equal, we get:

$$5s - 2(1-s) = -2(s) + 1(1-s)$$
$$5s - 2 + 2s = -2s + 1 - s$$
$$7s - 2 = -3s + 1$$
$$7s + 3s - 2 = -3s + 3s + 1$$
$$10s - 2 = 1$$
$$10s - 2 + 2 = 1 + 2$$
$$10s = 3$$
$$\frac{10s}{10} = \frac{3}{10}$$
$$s = \frac{3}{10}$$

So Player B should play her black card $\frac{3}{10}$ of the time and her white card $\frac{7}{10}$ of the time. Player B's payout when $s = \frac{3}{10}$ is

$$
\begin{aligned}
\$5s - \$2(1 - s) &= 5\left(\frac{3}{10}\right) - \$2\left(\frac{7}{10}\right) \\
&= \frac{15}{10} - \frac{14}{10} \\
&= \frac{1}{10} \text{ or } \$0.10
\end{aligned}
$$

Hence the maximum amount that Player B will lose on average is \$0.10 per game no matter what Player A does.

When both players use their strategy, the results can be shown by combining the two tree diagrams and calculating Player A's expected gain as shown in Figure 11-3.

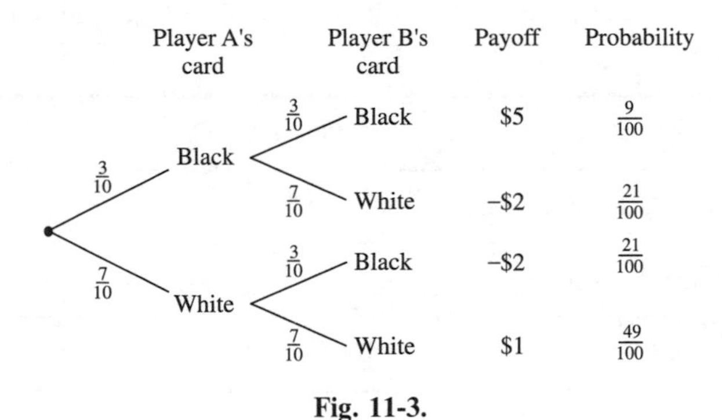

Fig. 11-3.

Hence, Player A's expected gain is

$$
\$5\left(\frac{9}{100}\right) - \$2\left(\frac{21}{100}\right) - \$2\left(\frac{21}{100}\right) + \$1\left(\frac{49}{100}\right)
$$

$$
\frac{45}{100} - \frac{42}{100} - \frac{42}{100} + \frac{49}{100} = \frac{10}{100} = \$0.10
$$

The number \$0.10 is called the **value** of the game. If the value of the game is 0, then the game is said to be fair.

The *optimal strategy* for Player A is to play the black card $\frac{3}{10}$ of the time and the white card $\frac{7}{10}$ of the time. The optimal strategy for Player B is the same in this case.

The **optimal strategy** for Player A is defined as a strategy that can guarantee him an average payoff of V(the value of the game) no matter what strategy Player B uses. The **optimal strategy** for Player B is defined as a strategy that prevents Player A from obtaining an average payoff greater than V(the value of the game) no matter what strategy Player A uses.

Note: When a player selects one strategy some of the time and another strategy at other times, it is called a *mixed* strategy, as opposed to using the same strategy all of the time. When the same strategy is used all of the time, it is called a *pure* strategy.

EXAMPLE: Two generals, A and B, decide to play a game. General A can attack General B's city either by land or by sea. General B can defend either by land or sea. They agree on the following payoff.

		General B (defend)	
		Land	Sea
General A (attack)	Land	−$25	$75
	Sea	$90	−$50

Find the optimal strategy for each player and the value of the game.

SOLUTION:

Let $p =$ probability of attacking by land and $1 - p =$ the probability of attacking by sea.

General A's expected payoff if he attacks and General B defends by land is $-\$25p + \$90(1-p)$ and if General B defends by sea is $\$75p - \$50(1-p)$. Equating the two and solving for p, we get

$$-25p + 90(1-p) = 75p - 50(1-p)$$
$$-25p + 90 - 90p = 75p - 50 + 50p$$
$$-115p + 90 = 125p - 50$$
$$-115p - 125p + 90 = 125p - 125p - 50$$
$$-240p + 90 = -50$$
$$-240p + 90 - 90 = -50 - 90$$
$$-240p = -140$$
$$\frac{-240p}{-240} = \frac{-140}{-240}$$
$$p = \frac{7}{12}$$

Hence General A should attack by land $\frac{7}{12}$ of the time and by sea $\frac{5}{12}$ of the time. The value of the game is

$$-25p + 90(1-p)$$
$$= -25\left(\frac{7}{12}\right) + 90\left(\frac{5}{12}\right)$$
$$= \$22.92$$

Now let us figure out General B's strategy.

General B should defend by land with a probability of s and by sea with a probability of $1-s$. Hence,

$$-25s + 75(1 - s) = 90s - 50(1 - s)$$

$$-25s + 75 - 75s = 90s - 50 + 50s$$

$$-100s + 75 = 140s - 50$$

$$-100s - 140s + 75 = 140s - 140s - 50$$

$$-240s + 75 = -50$$

$$-240s + 75 - 75 = -50 - 75$$

$$-240s = -125$$

$$\frac{-240s}{-240} = \frac{-125}{-240}$$

$$s = \frac{25}{48}$$

Hence, General B should defend by land with a probability of $\frac{25}{48}$ and by sea with a probability of $\frac{23}{48}$. A tree diagram for this problem is shown in Figure 11-4.

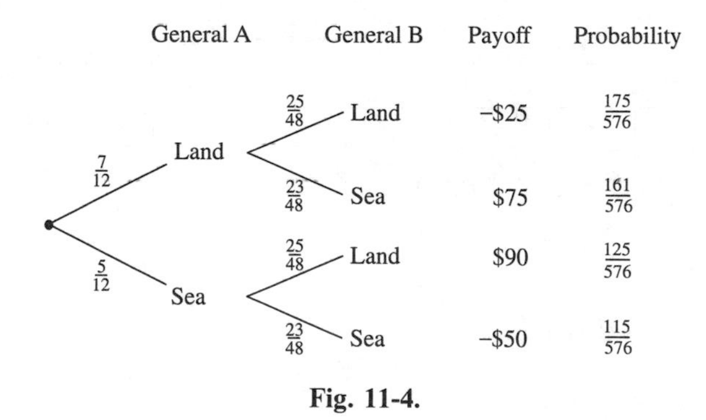

Fig. 11-4.

A payoff table can also consist of probabilities. This type of problem is shown in the next example.

EXAMPLE: Player A and Player B decided to play one-on-one basketball. Player A can take either a long shot or a lay-up shot. Player B can defend against either one. The payoff table shows the probabilities of a successful shot for each situation. Find the optimal strategy for each player and the value of the game.

	Player B (defense)	
Player A (offense)	Long shot	Lay-up shot
Long shot	.1	.4
Lay-up shot	.7	.2

SOLUTION:

Let p be the probability of shooting a long shot and $1-p$ the probability of shooting a lay-up shot. Then the probability of making a shot against a long shot defense is $0.1p + 0.7(1-p)$ and against a lay-up defense is $0.4p + 0.2(1-p)$. Equating and solving for p we get

$$0.1p + 0.7(1-p) = 0.4p + 0.2(1-p)$$

$$0.1p + 0.7 - 0.7p = 0.4p + 0.2 - 0.2p$$

$$-0.6p + 0.7 = 0.2p + 0.2$$

$$-0.6p + 0.7 - 0.2p = 0.2p + 0.2 - 0.2p$$

$$-0.8p + 0.7 = 0.2$$

$$-0.8p + 0.7 - 0.7 = 0.2 - 0.7$$

$$-0.8p = -0.5$$

$$\frac{-0.8p}{-0.8} = \frac{-0.5}{-0.8}$$

$$p = \frac{5}{8}$$

Then the probability of making a successful shot when $p = \frac{5}{8}$ is

$$\begin{aligned} 0.1p + 0.7(1-p) &= 0.1\left(\frac{5}{8}\right) + 0.7\left(1 - \frac{5}{8}\right) \\ &= 0.1\left(\frac{5}{8}\right) + 0.7\left(\frac{3}{8}\right) \\ &= \frac{13}{40} \end{aligned}$$

Hence Player A will be successful $\frac{13}{40}$ of the time. The value of the game is $\frac{13}{40}$.

Player B should defend against a long shot $\frac{1}{4}$ of the time and against a lay-up shot $\frac{3}{4}$ of the time. The solution is left as a practice question. See Question 5.

PRACTICE

1. A simplified version of football can be thought of as two types of plays. The offense can either run or pass and the defense can defend against a running play or a passing play. The payoff yards gained for each play are shown in the payoff box. Find the optimal strategy for each and determine the value of the game.

	Defense	
Offense	Against Run	Against Pass
Run	2	5
Pass	10	−6

2. In a game of paintball, a player can either hide behind a rock or in a tree. The other player can either select a pistol or a rifle. The probabilities for success are given in the payoff box. Determine the optimal strategy and the value of the game.

	Player B	
Player A	Rock	Tree
Pistol	0.5	0.2
Rifle	0.3	0.8

3. Person A has two cards, an ace (one) and a three. Person B has two cards, a two and a four. Each person plays one card. If the sum of the cards is 3 or 7, Person B pays Person A \$3 or \$7 respectively, but if the sum of the cards is 5, Person A pays Person B \$5. Construct a payoff table, determine the optimal strategy for each player, and the value of the game. Is the game fair?
4. A street vendor without a license has a choice to open on Main Street or Railroad Avenue. The city inspector can only visit one location per day. If he catches the vendor, the vendor must pay a \$50 fine; otherwise, the vendor can make \$100 at Main Street or \$75 at Railroad Avenue. Construct the payoff table, determine the optimal strategy for both locations, and find the value of the game.
5. Find the optimal strategy for Player B in the last example (basketball).

ANSWERS

1. Let p be the probability of running and $1-p=$ the probability of passing. Then

$$2p + 10(1 - p) = 5p - 6(1 - p)$$

$$2p + 10 - 10p = 5p - 6 + 6p$$

$$-8p + 10 = 11p - 6$$

$$-8p - 11p + 10 = 11p - 11p - 6$$

$$-19p + 10 = -6$$

$$-19p + 10 - 10 = -6 - 10$$

$$-19p = -16$$

$$\frac{-19p}{-19} = \frac{-16}{-19}$$

$$p = \frac{16}{19}$$

Hence, the player (offense) should run $\frac{16}{19}$ of the time, and pass $\frac{3}{19}$ of the time.

The value of the game when $p = \frac{16}{19}$ is

$$2p + 10(1-p) = 2\left(\frac{16}{19}\right) + 10\left(1 - \frac{16}{19}\right)$$

$$= 2\left(\frac{16}{19}\right) + 10\left(\frac{3}{19}\right)$$

$$= 3\frac{5}{19}$$

Let $s =$ the probability of defending against the run and $1 - s =$ the probability of defending against the pass; then

$$2s + 5(1-s) = 10s - 6(1-s)$$

$$2s + 5 - 5s = 10s - 6 + 6s$$

$$-3s + 5 = 16s - 6$$

$$-3s - 16s + 5 = 16s - 16s - 6$$

$$-19s + 5 = -6$$

$$-19s + 5 - 5 = -6 - 5$$

$$-19s = -11$$

$$\frac{-19s}{-19} = \frac{-11}{-19}$$

$$s = \frac{11}{19}$$

Hence, the player (defence) should defend against the run $\frac{11}{19}$ of the time.

2. Let p be the probability of selecting a pistol and $1-p$ be the probability of selecting a rifle.

$$0.5p + 0.3(1-p) = 0.2p + 0.8(1-p)$$
$$0.5p + 0.3 - 0.3p = 0.2p + 0.8 - 0.8p$$
$$0.2p + 0.3 = -0.6p + 0.8$$
$$0.2p + 0.6p + 0.3 = -0.6p + 0.6p + 0.8$$
$$0.8p + 0.3 = 0.8$$
$$0.8p + 0.3 - 0.3 = 0.8 - 0.3$$
$$0.8p = 0.5$$
$$\frac{0.8p}{0.8} = \frac{0.5}{0.8}$$
$$p = \frac{5}{8}$$

The value of the game when $p = \frac{5}{8}$ is

$$\begin{aligned} 0.5p + 0.3(1-p) &= 0.5\left(\frac{5}{8}\right) + 0.3\left(1 - \frac{5}{8}\right) \\ &= 0.5\left(\frac{5}{8}\right) + 0.3\left(\frac{3}{8}\right) \\ &= \frac{17}{40} \end{aligned}$$

When Player A selects the pistol $\frac{5}{8}$ of the time, he will be successful $\frac{17}{40}$ of the time.

Let s = the probability of Player B hiding behind a rock and $1 - s$ = the probability of hiding in a tree; then

$$0.5s + 0.2(1 - s) = 0.3s + 0.8(1 - s)$$

$$0.5s + 0.2 - 0.2s = 0.3s + 0.8 - 0.8s$$

$$0.3s + 0.2 = -0.5s + 0.8$$

$$0.3s + 0.5s + 0.2 = -0.5s + 0.5s + 0.8$$

$$0.8s + 0.2 = 0.8$$

$$0.8s + 0.2 - 0.2 = 0.8 - 0.2$$

$$0.8s = 0.6$$

$$\frac{0.8s}{0.8} = \frac{0.6}{0.8}$$

$$s = \frac{3}{4}$$

Hence, player B should hide behind the rock 3 times out of 4.

3. The payoff table is

	Player B	
Player A	Two	Four
Ace	3	–5
Three	–5	7

Let p be the probability that Player A plays the ace and $1-p$ be the probability that Player A plays the three. Then

$$\begin{aligned}
3p - 5(1-p) &= -5p + 7(1-p) \\
3p - 5 + 5p &= -5p + 7 - 7p \\
8p - 5 &= -12p + 7 \\
8p + 12p - 5 &= -12p + 12p + 7 \\
20p - 5 + 5 &= 7 + 5 \\
20p &= 12 \\
p &= \frac{12}{20} = \frac{3}{5}
\end{aligned}$$

The value of the game when $p = \frac{3}{5}$ is

$$\begin{aligned}
3p - 5(1-p) &= 3\left(\frac{3}{5}\right) - 5\left(1 - \frac{3}{5}\right) \\
&= 3\left(\frac{3}{5}\right) - 5\left(\frac{2}{5}\right) \\
&= -\frac{1}{5} \text{ or } -\$0.20
\end{aligned}$$

Player A will lose on average \$0.20 per game. Thus, the game is not fair.

Let s be the probability that Player B plays the two and $1-s$ be the probability that Player B plays the four; then

$$\begin{aligned}
3s - 5(1-s) &= -5s + 7(1-s) \\
3s - 5 + 5s &= -5s + 7 - 7s \\
8s - 5 &= -12s + 7 \\
8s + 12s - 5 &= -12s + 12s + 7 \\
20s - 5 &= 7 \\
20s - 5 + 5 &= 7 + 5 \\
20s &= 12 \\
\frac{20s}{20} &= \frac{12}{20} \\
s &= \frac{12}{20} = \frac{3}{5}
\end{aligned}$$

Player B should play the two, 3 times out of 5.

4. The payoff table is

	Inspector:	
Vendor:	Main St.	Railroad Ave.
Main St.	–\$50	\$100
Railroad Ave.	\$75	–\$50

Let p be the probability that the vendor selects Main St. and $1-p$ be the probability that the vendor selects Railroad Ave. Then,

$$-50p + 75(1-p) = 100p - 50(1-p)$$

$$-50p + 75 - 75p = 100p - 50 + 50p$$

$$-125p + 75 = 150p - 50$$

$$-125p - 150p + 75 = 150p - 150p - 50$$

$$-275p + 75 = -50$$

$$-275p + 75 - 75 = -50 - 75$$

$$-275p = -125$$

$$\frac{-275p}{-275} = \frac{-125}{-275}$$

$$p = \frac{125}{275} = \frac{5}{11}$$

The value of the game when $p = \frac{5}{11}$ is

$$-50p + 75(1-p) = -50\left(\frac{5}{11}\right) + 75\left(1 - \frac{5}{11}\right)$$

$$= -50\left(\frac{5}{11}\right) + 75\left(\frac{6}{11}\right)$$

$$= 18\frac{2}{11} \approx 18.18$$

Thus, if the vendor selects Main St. 5 times out of 11, he will make $\$18\frac{2}{11}$.

Let s be the probability that the inspector shows up at Main St. and $1-s$ be the probability that the inspector shows up at Railroad Ave. Then

$$-50s + 100(1-s) = 75s - 50(1-s)$$

$$-50s + 100 - 100s = 75s - 50 + 50s$$

$$-150s + 100 = 125s - 50$$

$$-150s - 125s + 100 = 125s - 125s - 50$$

$$-275s + 100 = -50$$

$$-275s + 100 - 100 = -50 - 100$$

$$-275s = -150$$

$$\frac{-275s}{-275s} = \frac{-150}{-275}$$

$$s = \frac{6}{11}$$

The inspector's optimal strategy will be to show up at the Main St. 6 times out of 11.

5. Let s be the probability that Player B defends against the long shot and $1-s$ be the probability that Player B defends against the lay-up. Then

$$0.1s + 0.4(1-s) = 0.7s + 0.2(1-s)$$

$$0.1s + 0.4 - 0.4s = 0.7s + 0.2 - 0.2s$$

$$-0.3s + 0.4 = 0.5s + 0.2$$

$$-0.3s - 0.5s + 0.4 = 0.5s - 0.5s + 0.2$$

$$-0.8s + 0.4 = 0.2$$

$$-0.8s + 0.4 - 0.4 = 0.2 - 0.4$$

$$-0.8s = -0.2$$

$$\frac{-0.8s}{-0.8} = \frac{-0.2}{-0.8}$$

$$s = \frac{1}{4}$$

Hence, player B must defend against a long shot $\frac{1}{4}$ of the time, and against a lay-up shot $\frac{3}{4}$ of the time.

Summary

Game theory uses mathematics to analyze games. These games can range from simple board games to warfare. A game can be considered a contest between two players and consists of rules on how to play the game and how to determine the winner. In this chapter, only two-player, zero sum games were explained. A payoff table is used to determine how much a person wins or loses. Payoff tables can also consist of probabilities.

A strategy is a rule that determines a player's move or moves in order to win the game or maximize the player's payoff. An optimal strategy is the strategy that a player uses that will guarantee him or her an average payoff of a certain amount no matter what the other player does. An optimal strategy for a player could also be one that will prevent the other player from obtaining an average payoff greater that a certain amount. This amount is called the value of the game. If the value of the game is zero, then the game is fair.

CHAPTER QUIZ

1. The person who developed the concepts of game theory was
   a. Garry Kasparov
   b. Leonhard Euler
   c. John Von Neumann
   d. Blasé Pascal

2. The reward for winning the game is called the

   a. Bet
   b. Payoff
   c. Strategy
   d. Loss or win

3. In a game where one player pays the other player and vice versa, the game is called —— game.

   a. A payoff
   b. An even sum
   c. No win
   d. A zero sum

4. When both players use an optimal strategy, the amount that on average is the payoff over the long run is called the —— of the game.

   a. Value
   b. Winnings
   c. Strength
   d. Odds

5. If a game is fair, the value of the game will be

   a. 0
   b. 1
   c. −1
   d. Undetermined

Use the following payoff table to answer questions 6–10.

	Player B:	
Player A:	X	Y
X	2	5
Y	7	−8

6. If Player A chooses X and Player B chooses Y, the payoff is

   a. 2
   b. 7
   c. 5
   d. −8

7. The optimal strategy for Player A would be to select X with a probability of

   a. $\frac{1}{6}$

   b. $\frac{3}{8}$

   c. $\frac{5}{6}$

   d. $\frac{5}{8}$

8. When Player A plays X using his optimal strategy, the value of the game is

   a. $6\frac{1}{6}$

   b. $8\frac{3}{8}$

   c. $1\frac{3}{4}$

   d. $2\frac{5}{6}$

9. The optimal strategy for Player B would be to select Y with a probability of

   a. $\frac{13}{18}$

   b. $\frac{5}{18}$

   c. $\frac{2}{3}$

   d. $\frac{1}{3}$

10. When Player B uses his optimal strategy, the value of the game will be

    a. $3\frac{1}{6}$

    b. $5\frac{2}{3}$

    c. $2\frac{5}{6}$

    d. $4\frac{1}{8}$

Probability Sidelight

COMPUTERS AND GAME THEORY

Computers have been used to analyze games, most notably the game of chess. Experts have written programs enabling computers to play humans. Matches between chess champion Garry Kasparov and the computer named Deep Blue, as well as his matches against the newer computer X3D Fritz, have received universal notoriety.

Computers cannot think, but they can make billions of calculations per second. What the computer does when it is its turn to make a chess move is to generate a tree of moves. Each player has about 20 choices of a move per turn. Based on these choices, the computer calculates the possible moves of its human opponent; then it moves based on the human's possible moves. With each move, the computer evaluates the position of the chess pieces on the board at that time. Each chess piece is assigned a value based on its importance. For example, a pawn is worth one point, a knight is worth three points, a rook is worth five points, and a queen is worth nine points. The computer then works backwards, assuming its human opponent will make his best move. This process is repeated after each human move. It is not possible for the computer to make trees for an entire game since it has been estimated that there are 10^{1050} possible chess moves. By looking ahead several moves, the computer can play a fairly decent game. Some programs can beat almost all human opponents. (Chess champions excluded, of course!)

As the power of the computer increases, the more trees the computer will be able to evaluate within a specific time period. Also, computers have been able to be programmed to remember previous games, thus helping in its analysis of the trees. Many people think that in the future, a computer will be built that will be able to defeat all human players.

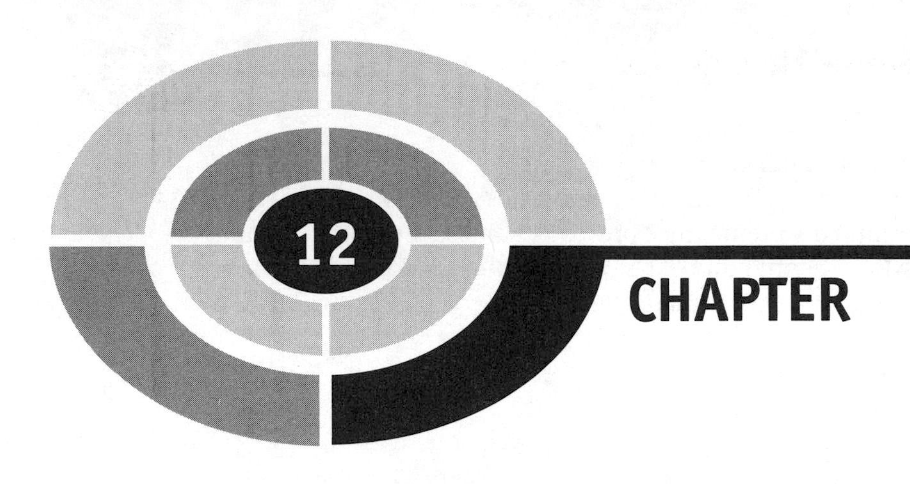

Actuarial Science

Introduction

An **actuary** is a person who uses mathematics to analyze risks in order to determine insurance rates, investment strategies, and other situations involving future payouts. Most actuaries work for insurance companies; however, some work for the United States government in the Social Security and Medicare programs and others as consultants to business and financial institutions. The main function of an actuary is to determine premiums for life and health insurance policies and retirement accounts, as well as premiums for flood insurance, mine subsidence, etc. Actuarial science involves several areas of mathematics, including calculus. However, much of actuarial science is based on probability.

Mortality Tables

Insurance companies collect data on various risk situations, such as life expectancy, automobile accidents, hurricane damages, etc. The information can be summarized in table form. One such table is called a *mortality table* or a *period life table*. You can find one at the end of this chapter. The **mortality table** used here is from the Social Security Administration and shows the ages for males and females, the probability of dying at a specific age, the number of males and females surviving during a specific year of their lives, and life expectancies for a given age. The following examples show how to use the mortality table.

EXAMPLE: Find the probability of a female dying during her 30th year.

SOLUTION:

Based on the mortality table, there are 98,428 females out of 100,000 alive at the beginning of year 30 and 98,366 females living at the beginning of year 31, so to find the number of females who have died during year 30, subtract $98{,}428 - 98{,}366 = 62$. Therefore, 62 out of 98,428 people have died. Next find the probability.

$$P(\text{dying at age 30}) = \frac{\text{number who died during the year}}{\text{number who were alive at the beginning of year 30}}$$

$$= \frac{62}{98{,}428} \approx 0.00063$$

(Notice that under the column labeled "Death probability," the figure given for 30-year-old females is 0.000624. The discrepancy is probably due to the fact that computations for this column were based on sample sizes larger than 100,000 or perhaps it was due to rounding.)

EXAMPLE: On average, how long can one expect a female who is 30 years old to live?

SOLUTION:

Looking at the table for 30-year-old females, the last column shows a life expectancy of 50.43 years. This means that at age 30, a female can expect to live on average another 50.43 years or to age $(30 + 50.43) = 80.43$ years. Interpreting this means that the average of the life expectancies of females age 30 is 50.43 years. Remember this is an **average,** not a guarantee.

EXAMPLE: Find the death rate for 30-year-old males.

SOLUTION:

From the table for a 30-year-old man, there are 97,129 out of 100,000 living, and for age 31, there are 96,999 males living; hence, 97,129–96,999, or 130 males died during their 30th year of life. Now the death rate is 130 males out of a total of 97,129 or

$$P(\text{dying at } 30) = \frac{\text{number who died during the year}}{\text{number living at the beginning of year 30}}$$

$$= \frac{130}{97{,}129} \approx 0.00133$$

Notice that the table gives a value of 0.001396 under the column "Death probability." The reason for this discrepancy is probably due to the fact that samples larger than 100,000 males were used in the calculation, or perhaps it is due to rounding.

EXAMPLE: What is the probability that a male age 25 will die before age 60?

SOLUTION:

The number of males living at age 25 is 97,760 out of 100,000, and the number of males living at age 60 is 84,682. So to find the number of males who died, subtract the two numbers: $97{,}760 - 84{,}682 = 13{,}078$. That is, 13,078 males died between age 25 and age 60. Next, find the probability.

$$P = \frac{\text{number who died}}{\text{number living at the beginning of year 25}}$$

$$= \frac{13{,}078}{97{,}760} \approx 0.134$$

In other words, there is about a 13% chance that a male age 25 will die before age 60.

EXAMPLE: What is the probability that a female who is 40 will live to the age of 60?

SOLUTION:

At age 40 there are 97,512 females out of 100,000 alive. At age 70, there are 79,880 females alive. Hence,

$$P(\text{live to } 70) = \frac{\text{number living at } 70}{\text{number living at } 40}$$

$$= \frac{79{,}880}{97{,}512} = 0.819$$

In other words, the probability of a 40-year-old female living to age 70 is 0.819, or about 82%.

EXAMPLE: How many males age 21 will die before age 65?

SOLUTION:

At age 21, there are 98,307 males out of 100,000 alive. At age 65, there are 78,410 males alive. Therefore, $98{,}307 - 78{,}410 = 19{,}897$ males have died between the ages of 21 and 65. This is out of 98,307 who made it to age 21.

PRACTICE

1. Find the probability that a male will die at age 48.
2. On average, how many more years can a female who is age 56 expect to live?
3. Find the death rate for a 63-year-old female.
4. What is the probability that a male age 43 will live to age 65?
5. What is the probability that a 25-year-old female will live to age 60?
6. Find the probability that a male will live to 21 years of age.
7. How many years longer can a female age 20 expect to live than a male age 20?
8. About how many 2-year-old males will die before they reach 10 years old?
9. What is the probability that a female age 16 will live to age 50?
10. Find the probability that a male will live to age 65.

ANSWERS

1. From the table, we see the probability that a male age 48 will die is 0.004859.

   Alternate Solution: There are 92,790 males out of 100,000 males age 48, and there are 92,339 males alive at age 49. So, $92{,}790 - 92{,}339 = 451$ males died at age 48.

$$P(\text{dying at age 48}) = \frac{\text{number who died at age 48}}{\text{number alive at age 48}} = \frac{451}{92{,}790} = 0.00486$$

2. From the table, a 56-year-old female can expect to live another 26.30 years.
3. From the table, the death rate for a 63-year-old female is 0.010598. Alternate Solution: At age 63, there are 88,518 out of 100,000 females alive, and there are 87,580 females alive at age 64. Hence, $88{,}518 - 87{,}580 = 938$ females died at age 63; then

$$P(\text{dying at age 63}) = \frac{\text{number who died at age 63}}{\text{number alive at age 63}} = \frac{938}{88{,}518} = 0.010597.$$

4. At age 43, there are 94,629 males out of 100,000 alive. At age 65, there are 78,410 males alive, so

$$P(\text{living to 65}) = \frac{\text{number alive at 65}}{\text{number alive at 43}} = \frac{78{,}410}{94{,}629} = 0.8286 \text{ or } 82.86\%$$

5. At 25, there are 98,689 females out of 100,000 alive. At age 60, there are 90,867 females alive; hence,

$$P(\text{living to age 60}) = \frac{\text{number alive at 60}}{\text{number alive at 25}} = \frac{90{,}867}{98{,}689} \approx 0.921 \text{ or } 92.1\%$$

6. At age 21, there are 98,307 out of 100,000 males alive; hence,

$$P(\text{a male will live to age 21}) = \frac{\text{number alive at 21}}{\text{total born}} = \frac{98{,}307}{100{,}000} = 0.98307$$

7. At age 20, a female can expect to live 60.16 more years. At age 20, a male can expect to live 55.04 more years. Hence, $60.16 - 55.04 = 5.12$. A female can expect to live 5.12 years longer than a male if both are age 20.
8. At age 2, there are 99,187 males out of 100,000 alive. At age 10, there are 99,013 males alive; hence, $99{,}187 - 99{,}013 = 174$ males age 2 who will die before age 10. This is out of 99,187 males alive at age 2.
9. There are 99,084 females out of 100,000 alive at age 16. There are 95,464 females alive at age 50. Hence,

$$P(\text{living to age 50}) = \frac{\text{number alive at 50}}{\text{number alive at 16}} = \frac{95{,}464}{99{,}084} = 0.963 = 96.3\%$$

10. There are 78,410 males out of 100,000 alive at age 65; hence,

$$P(\text{living to age 65}) = \frac{\text{number alive at 65}}{100{,}000} = \frac{78{,}410}{100{,}000} = 0.7841 = 78.41\%$$

Life Insurance Policies

There are many different types of life insurance policies. A **straight life insurance** policy requires that you make payments for your entire life. Then when you die, your *beneficiary* is paid the face value of the policy. A **beneficiary** is a person designated to receive the money from an insurance policy.

Another type of policy is a **term policy**. Here the insured pays a certain premium for twenty years. If the person dies during the 20-year period, his or her beneficiary receives the value of the policy. If the person lives beyond the twenty-year period, he or she receives nothing. This kind of insurance has low premiums, especially for younger people since the probability of them dying is relative small.

Another type of life insurance policy is called an **endowment** policy. In this case, if a person purchases a 20-year endowment policy and lives past 20 years, the insurance company will pay the face value of the policy to the insured. Naturally, the premiums for this kind of policy are much higher than those for a term policy.

The tables show the approximate premiums for a $100,000 20-year term policy. These are based on very healthy individuals. Insurance companies adjust the premiums for people with health problems.

Age	Male	Female
21	$115	$96
30	$147	$98
40	$151	$124

EXAMPLE: If a 21-year-old healthy female takes a 20-year term life insurance policy for $100,000, how much would she pay in premiums if she lived at least 20 years?

SOLUTION:

Her premium would be $96 per year, so she would pay $96 × 20 years = $1920.

EXAMPLE: If a healthy 30-year-old male takes a 20-year term life insurance policy for $25,000, how much would he pay if he lives for at least 20 years?

SOLUTION:

The premium for a healthy 30-year-old male for a 20-year term policy of $100,000 is $147. So for a $25,000 policy, the premium can be found by making a ratio equal to

$\frac{\text{face value of insurance policy}}{\$100{,}000}$ and multiplying it by the premium:

$\frac{\$25{,}000}{\$100{,}000} \times \$147 = \$36.75$. Then multiply by 20 years:

$\$36.75 \times 20 = \$735.$

EXAMPLE: If the life insurance company insures 100 healthy females age 40 for 20-year, $100,000 term life insurance policies, find the approximate amount the company will have to pay out.

SOLUTION:

First use the mortality table to find the probability that a female aged 40 will die before she reaches age 60. At age 40, there are 97,512 females out of 100,000 living. At age 60, there are 90,867 living. So, in twenty years, 97,512 − 90,867 = 6645 have died during the 20-year period. Hence, the probability of dying is

$$P(\text{dying}) = \frac{\text{number who have died}}{\text{number living at age 40}} = \frac{6645}{97{,}512} = 0.068$$

Hence about 6.8 or 7% (rounded) of the females have died during the 20-year period. If the company has insured 100 females, then about $7\% \times 100 = 7$ will die in the 20-year period. The company will have to pay out $7 \times \$100{,}000 = \$700{,}000$ in the 20-year period.

Notice that knowing this information, the insurance company can estimate its costs (overhead) and calculate premiums to determine its profit.

Another statistic that insurance companies use is called the *median future lifetime* of a group of individuals at a given age. The **median future lifetime** for people living at a certain age is the number of years that approximately one-half of those individuals will still be alive.

EXAMPLE: Find the median future lifetime for a male who is 30 years old.

SOLUTION:

Using the mortality table, find the number of males living at age 30. It is 97,129 out of 100,000. Then divide this number by 2 to get $97{,}129 \div 2 = 48{,}564.5$. Next, using the closest value, find the age of the males that corresponds to 48,564.5. That is 48,514. The age is 78. In other words, at age 78, about one-half of the males are still living. Subtract $78 - 30 = 48$. The median future lifetime of a 30-year-old male is 48 years.

PRACTICE

1. If a healthy 40-year-old male takes a 20-year, $100,000 term life insurance policy, how much would he pay in premiums if he lived to age 60?
2. If a healthy female age 21 takes a 20-year, $40,000 term life insurance policy, about how much would she pay in premiums if she lived to age 41?
3. If a life insurance company insures 100 healthy females age 35 for $50,000, 20-year term policies, how much would they expect to pay out?
4. Find the median future lifetime of a female who is age 35.
5. Find the median future lifetime of a male who is age 50.

ANSWERS

1. $\$151 \times 20 = \$3020$

2. $\dfrac{\$40{,}000}{\$100{,}000} \times 96 \times 20 = \$768$

3. At age 35, there are 98,067 females out of 100,000 alive. At age 55, there are 93,672 females alive. Therefore, $98{,}067 - 93{,}672 = 4395$ females will die.

   $$P(\text{dying in 20 years}) = \frac{\text{number who will die}}{\text{number alive at 35}} = \frac{4395}{98{,}067} = 0.0448$$

   Out of 100 females, $100 \times 0.0448 = 4.48$ or about 5 will die. Hence, $5 \times \$100{,}000 = \$500{,}000$ will have to be paid out.

4. At age 35, there are 98,067 females out of 100,000 alive; $98{,}067 \div 2 = 49{,}033.5$. At age 83, there are 48,848 females alive. So, $83 - 35 = 48$ is the median future lifetime.

5. At age 50, there are 91,865 males out of 100,000 alive; $91{,}865 \div 2 = 45{,}932.5$. At age 79, there are 45,459 males alive. Hence, the median future lifetime of a male age 50 is $79 - 50 = 29$ years.

Summary

This chapter introduces some of the concepts used in actuarial science. An actuary is a person who uses mathematics in order to determine insurance rates, investment strategies, retirement accounts and other situations involving future payouts.

Actuaries use mortality tables to determine the probabilities of people living to certain ages. A mortality table shows the number of people out of 1,000, 10,000, or 100,000 living at certain ages. It can also show the probability of dying at any given age. Barring unforeseen catastrophic events such as wars, plagues, and such, the number of people dying at a specific age is relatively constant for certain groups of people.

In addition to life insurance, mortality tables are used in other areas. Some of these include Social Security and retirement accounts.

CHAPTER QUIZ

1. The probability of a male age 33 dying before age 48 is
   a. 0.96
   b. 0.33
   c. 0.72
   d. 0.04
2. The probability that a female age 72 will die is
   a. 0.052
   b. 0.024
   c. 0.037
   d. 0.041
3. The life expectancy of a female who is 47 is
   a. 34.34 years
   b. 23.26 years
   c. 15.93 years
   d. 9.87 years
4. The probability that a male age 28 will live to age 56 is
   a. 0.094
   b. 0.873
   c. 0.906
   d. 0.127
5. The probability that a female age 26 will live until age 77 is
   a. 0.329
   b. 0.527
   c. 0.671
   d. 0.473
6. The probability that a female will live to age 50 is
   a. 0.955
   b. 0.045
   c. 0.191
   d. 0.081

7. How much will a healthy 40-year-old female pay for a $100,000, 20-year term policy if she lives to age 60?

   a. $3920
   b. $5880
   c. $6040
   d. $2480

8. If a life insurance company writes 100 males age 21 a $30,000, 20-year term policy, how much will it pay out in 20 years?

   a. $90,000
   b. $120,000
   c. $60,000
   d. $150,000

9. The median future lifetime of a 51-year-old male is

   a. 15 years
   b. 32 years
   c. 30 years
   d. 28 years

10. The median future lifetime of a 63-year-old female is

   a. 27 years
   b. 21 years
   c. 19 years
   d. 30 years

Probability Sidelight

EARLY HISTORY OF MORTALITY TABLES

Surveys and censuses have been around for a long time. Early rulers wanted to keep track of the economic wealth and manpower of their subjects. One of the earliest enumeration records appears in the Bible in the Book of Numbers. Egyptian and Roman rulers were noted for their surveys and censuses.

In the late 1500s and early 1600s, parish clerks of the Church of England in London began keeping records of the births, deaths, marriages, and baptisms of their parishioners. Many of these were published weekly and summarized yearly. They were called the *Bills of Mortality*. Some even included possible causes of death as well as could be determined at that time. At best, they were "hit and miss" accounts. If a clerk did not publish the information one week, the figures were included in the next week's summary. Also during this time, people began keeping records of deaths due to the various plagues.

Around 1662, an English merchant, John Graunt (1620–1674), began reviewing the *Bills of Mortality* and combining them into tables. He used records from the years 1604 to 1661 and produced tables that he published in a book entitled *National and Political Observations.* He noticed that with the exception of plagues or wars, the number of people that died at a certain age was fairly consistent. He then produced a crude mortality table from this information. After reviewing the data, he drew several conclusions. Some were accurate and some were not.

He stated that the number of male births was slightly greater than the number of female births. He also noticed that, in general, women lived longer than men. He stated that physicians treated about twice as many female patients as male patients, and that they were better able to cure the female patients. From this fact, he concluded that either men were more prone to die from their vices or that men didn't go to the doctor as often as women when they were ill!

For his work in this area, he was given a fellowship in the Royal Society of London. He was the first merchant to receive this honor. Until this time, all members were doctors, noblemen, and lawyers.

Two brothers from Holland, Ludwig and Christiaan Huygens (1629–1695) noticed his work. They expanded on Gaunt's work and constructed their own mortality table. This was the first table that used probability theory and included the probabilities of a person dying at a certain age in his or her life and also the probability of surviving to a certain age.

Later, insurance companies began producing and using mortality tables to determine life expectancies and rates for life insurance.

Period Life Table, 2001 (Updated June 16, 2004)

Exact age	Male			Female		
	Death probability[1]	Number of lives[2]	Life expectancy	Death probability[1]	Number of lives[2]	Life expectancy
0	0.007589	100,000	73.98	0.006234	100,000	79.35
1	0.000543	99,241	73.54	0.000447	99,377	78.84
2	0.000376	99,187	72.58	0.000301	99,332	77.88
3	0.000283	99,150	71.61	0.000198	99,302	76.90
4	0.000218	99,122	70.63	0.000188	99,283	75.92
5	0.000199	99,100	69.64	0.000165	99,264	74.93
6	0.000191	99,081	68.66	0.000150	99,248	73.94
7	0.000183	99,062	67.67	0.000139	99,233	72.95
8	0.000166	99,043	66.68	0.000129	99,219	71.96
9	0.000144	99,027	65.69	0.000120	99,206	70.97
10	0.000126	99,013	64.70	0.000115	99,194	69.98
11	0.000133	99,000	63.71	0.000120	99,183	68.99
12	0.000189	98,987	62.72	0.000142	99,171	68.00
13	0.000305	98,968	61.73	0.000184	99,157	67.01
14	0.000466	98,938	60.75	0.000241	99,139	66.02
15	0.000642	98,892	59.78	0.000305	99,115	65.04
16	0.000808	98,829	58.81	0.000366	99,084	64.06
17	0.000957	98,749	57.86	0.000412	99,048	63.08

(*Continued*)

Continued

Exact age	Male			Female		
	Death probability[1]	Number of lives[2]	Life expectancy	Death probability[1]	Number of lives[2]	Life expectancy
18	0.001078	98,654	56.92	0.000436	99,007	62.10
19	0.001174	98,548	55.98	0.000444	98,964	61.13
20	0.001271	98,432	55.04	0.000450	98,920	60.16
21	0.001363	98,307	54.11	0.000460	98,876	59.19
22	0.001415	98,173	53.19	0.000468	98,830	58.21
23	0.001415	98,034	52.26	0.000475	98,784	57.24
24	0.001380	97,896	51.33	0.000484	98,737	56.27
25	0.001330	97,760	50.40	0.000492	98,689	55.29
26	0.001291	97,630	49.47	0.000504	98,641	54.32
27	0.001269	97,504	48.53	0.000523	98,591	53.35
28	0.001275	97,381	47.59	0.000549	98,539	52.38
29	0.001306	97,256	46.65	0.000584	98,485	51.40
30	0.001346	97,129	45.72	0.000624	98,428	50.43
31	0.001391	96,999	44.78	0.000670	98,366	49.46
32	0.001455	96,864	43.84	0.000724	98,301	48.50
33	0.001538	96,723	42.90	0.000788	98,229	47.53
34	0.001641	96,574	41.97	0.000862	98,152	46.57
35	0.001761	96,416	41.03	0.000943	98,067	45.61

(Continued)

Continued

Exact age	Male			Female		
	Death probability[1]	Number of lives[2]	Life expectancy	Death probability[1]	Number of lives[2]	Life expectancy
36	0.001895	96,246	40.11	0.001031	97,975	44.65
37	0.002044	96,063	39.18	0.001127	97,874	43.70
38	0.002207	95,867	38.26	0.001231	97,764	42.75
39	0.002385	95,656	37.34	0.001342	97,643	41.80
40	0.002578	95,427	36.43	0.001465	97,512	40.85
41	0.002789	95,181	35.52	0.001597	97,369	39.91
42	0.003025	94,916	34.62	0.001730	97,214	38.98
43	0.003289	94,629	33.73	0.001861	97,046	38.04
44	0.003577	94,318	32.84	0.001995	96,865	37.11
45	0.003902	93,980	31.95	0.002145	96,672	36.19
46	0.004244	93,613	31.08	0.002315	96,464	35.26
47	0.004568	93,216	30.21	0.002498	96,241	34.34
48	0.004859	92,790	29.34	0.002693	96,001	33.43
49	0.005142	92,339	28.48	0.002908	95,742	32.52
50	0.005450	91,865	27.63	0.003149	95,464	31.61
51	0.005821	91,364	26.78	0.003424	95,163	30.71
52	0.006270	90,832	25.93	0.003739	94,837	29.81
53	0.006817	90,263	25.09	0.004099	94,483	28.92

(*Continued*)

Continued

Exact age	Male			Female		
	Death probability[1]	Number of lives[2]	Life expectancy	Death probability[1]	Number of lives[2]	Life expectancy
54	0.007457	89,647	24.26	0.004505	94,095	28.04
55	0.008191	88,979	23.44	0.004969	93,672	27.16
56	0.008991	88,250	22.63	0.005482	93,206	26.30
57	0.009823	87,457	21.83	0.006028	92,695	25.44
58	0.010671	86,597	21.04	0.006601	92,136	24.59
59	0.011571	85,673	20.26	0.007220	91,528	23.75
60	0.012547	84,682	19.49	0.007888	90,867	22.92
61	0.013673	83,620	18.73	0.008647	90,151	22.10
62	0.015020	82,476	17.99	0.009542	89,371	21.29
63	0.016636	81,237	17.25	0.010598	88,518	20.49
64	0.018482	79,886	16.54	0.011795	87,580	19.70
65	0.020548	78,410	15.84	0.013148	86,547	18.93
66	0.022728	76,798	15.16	0.014574	85,409	18.18
67	0.024913	75,053	14.50	0.015965	84,164	17.44
68	0.027044	73,183	13.86	0.017267	82,821	16.71
69	0.029211	71,204	13.23	0.018565	81,391	16.00
70	0.031632	69,124	12.61	0.020038	79,880	15.29
71	0.034378	66,937	12.01	0.021767	78,279	14.59

(*Continued*)

Continued

Exact age	Male			Female		
	Death probability[1]	Number of lives[2]	Life expectancy	Death probability[1]	Number of lives[2]	Life expectancy
72	0.037344	64,636	11.42	0.023691	76,575	13.91
73	0.040545	62,223	10.84	0.025838	74,761	13.23
74	0.044058	59,700	10.28	0.028258	72,829	12.57
75	0.048038	57,069	9.73	0.031076	70,771	11.92
76	0.052535	54,328	9.20	0.034298	68,572	11.29
77	0.057503	51,474	8.68	0.037847	66,220	10.67
78	0.062971	48,514	8.18	0.041727	63,714	10.07
79	0.069030	45,459	7.69	0.046048	61,055	9.49
80	0.075763	42,321	7.23	0.051019	58,244	8.92
81	0.083294	39,115	6.78	0.056721	55,272	8.37
82	0.091719	35,857	6.35	0.063095	52,137	7.85
83	0.101116	32,568	5.94	0.070179	48,848	7.34
84	0.111477	29,275	5.55	0.078074	45,420	6.86
85	0.122763	26,011	5.18	0.086900	41,873	6.39
86	0.134943	22,818	4.84	0.096760	38,235	5.96
87	0.148004	19,739	4.52	0.107728	34,535	5.54
88	0.161948	16,817	4.21	0.119852	30,815	5.15
89	0.176798	14,094	3.93	0.133149	27,121	4.78

(*Continued*)

Continued

Exact age	Male			Female		
	Death probability[1]	Number of lives[2]	Life expectancy	Death probability[1]	Number of lives[2]	Life expectancy
90	0.192573	11,602	3.67	0.147622	23,510	4.44
91	0.209287	9,368	3.42	0.163263	20,040	4.12
92	0.226948	7,407	3.20	0.180052	16,768	3.83
93	0.245551	5,726	2.99	0.197963	13,749	3.56
94	0.265081	4,320	2.80	0.216961	11,027	3.31
95	0.284598	3,175	2.63	0.236221	8635	3.09
96	0.303872	2,271	2.47	0.255493	6595	2.89
97	0.322655	1,581	2.34	0.274498	4910	2.71
98	0.340694	1,071	2.21	0.292942	3562	2.55
99	0.357729	706	2.10	0.310519	2519	2.40
100	0.375615	454	1.98	0.329150	1737	2.26
101	0.394396	283	1.88	0.348899	1165	2.12
102	0.414116	171	1.78	0.369833	759	1.99
103	0.434821	100	1.68	0.392023	478	1.87
104	0.456562	57	1.58	0.415544	291	1.75
105	0.479391	31	1.49	0.440477	170	1.63
106	0.503360	16	1.40	0.466905	95	1.52
107	0.528528	8	1.32	0.494920	51	1.42

(*Continued*)

Continued

	Male			Female		
Exact age	Death probability[1]	Number of lives[2]	Life expectancy	Death probability[1]	Number of lives[2]	Life expectancy
108	0.554954	4	1.24	0.524615	26	1.32
109	0.582702	2	1.16	0.556092	12	1.23
110	0.611837	1	1.09	0.589457	5	1.14
111	0.642429	0	1.02	0.624824	2	1.05
112	0.674551	0	0.95	0.662314	1	0.97
113	0.708278	0	0.89	0.702053	0	0.89
114	0.743692	0	0.82	0.743692	0	0.82
115	0.780876	0	0.76	0.780876	0	0.76
116	0.819920	0	0.71	0.819920	0	0.71
117	0.860916	0	0.65	0.860916	0	0.65
118	0.903962	0	0.60	0.903962	0	0.60
119	0.949160	0	0.55	0.949160	0	0.55

[1]Probability of dying within one year.
[2]Number of survivors out of 100,000 born alive.

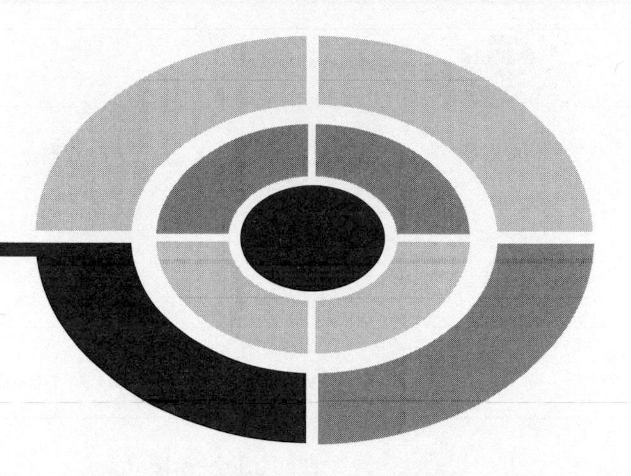

Final Exam

Select the best answer.

1. The list of all possible outcomes of a probability experiment is called the

   a. Experimental space
   b. Probability space
   c. Random space
   d. Sample space

2. The probability of an event can be any number from _____ to _____.

   a. −1, 1
   b. 0, 1
   c. 1, 100
   d. 0, infinity

3. If an event cannot occur, its probability is
   a. 0
   b. $\frac{1}{2}$
   c. 1
   d. -1

4. When two events cannot occur at the same time, they are said to be _____ events.
   a. Independent
   b. Mutually exclusive
   c. Random
   d. Inconsistent

5. If the probability that an event will happen is 0.48, then the probability that the event will not happen is
   a. 0.52
   b. -0.48
   c. 0
   d. 1

6. When two dice are rolled, the sample space consists of _____ outcomes.
   a. 6
   b. 12
   c. 18
   d. 36

7. What is 0!?
   a. 0
   b. 1
   c. Undefined
   d. Infinite

8. The sum of the probabilities of each outcome in the sample space will always be
   a. 0
   b. $\frac{1}{2}$
   c. 1
   d. Different

9. When two dice are rolled, the probability of getting a sum of 8 is
   a. $\frac{5}{36}$
   b. $\frac{1}{6}$
   c. $\frac{8}{36}$
   d. 0
10. A Gallup poll found that 78% of Americans worry about the quality and healthfulness of their diets. If three people are selected at random, the probability that all three will worry about the healthfulness and quality of their diets is
   a. 2.34
   b. 0.78
   c. 0.22
   d. 0.47
11. When a die is rolled, the probability of getting a number less than 5 is
   a. $\frac{5}{36}$
   b. $\frac{1}{3}$
   c. $\frac{2}{3}$
   d. $\frac{5}{6}$
12. When a card is drawn from a deck of 52 cards, the probability of getting a heart is
   a. $\frac{1}{13}$
   b. $\frac{1}{2}$
   c. $\frac{1}{4}$
   d. $\frac{5}{52}$

13. When a die is rolled, the probability of getting an odd number less than three is

   a. $\frac{1}{6}$

   b. 0

   c. $\frac{1}{3}$

   d. $\frac{1}{2}$

14. A survey conducted at a local restaurant found that 18 people preferred orange juice, 12 people preferred grapefruit juice, and 6 people preferred apple juice with their breakfasts. If a person is selected at random, the probability that the person will select apple juice is

   a. $\frac{1}{2}$

   b. $\frac{1}{3}$

   c. $\frac{1}{4}$

   d. $\frac{1}{6}$

15. During a sale at a men's store, 16 white sweaters, 3 red sweaters, 9 blue sweaters, and 7 yellow sweaters were purchased. If a customer is selected at random, find the probability that the customer purchased a yellow or a white sweater.

   a. $\frac{23}{35}$

   b. $\frac{9}{35}$

   c. $\frac{19}{35}$

   d. $\frac{7}{35}$

16. A card is selected from an ordinary deck of 52 cards. The probability that it is a 7 or a heart is
    a. $\frac{17}{52}$
    b. $\frac{1}{4}$
    c. $\frac{4}{13}$
    d. $\frac{1}{13}$

17. Two dice are rolled; the probability of getting a sum greater than or equal to 9 is
    a. $\frac{1}{6}$
    b. $\frac{5}{18}$
    c. $\frac{1}{4}$
    d. $\frac{1}{9}$

18. A card is selected from an ordinary deck of 52 cards. The probability that it is a red face card is
    a. $\frac{3}{26}$
    b. $\frac{3}{13}$
    c. $\frac{1}{2}$
    d. $\frac{2}{13}$

19. Three cards are drawn from an ordinary deck of 52 cards without replacement. The probability of getting three queens is

a. $\frac{1}{2197}$

b. $\frac{3}{52}$

c. $\frac{1}{5525}$

d. $\frac{1}{169}$

20. An automobile license plate consists of 3 letters followed by 2 digits. The number of different plates that can be made if repetitions are not permitted is

a. 7800
b. 1,404,000
c. 1,757,600
d. 6318

21. The number of different arrangements of the letters of the word *next* is

a. 256
b. 24
c. 18
d. 16

22. A psychology quiz consists of 12 true-false questions. The number of possible different answer keys that can be made is

a. 24
b. 144
c. 47,900,600
d. 4096

23. How many different ways can 4 books be selected from 7 books?

a. 210
b. 35
c. 28
d. 840

24. The number of different ways 5 boys and 4 girls can be selected from 7 boys and 9 girls is

    a. 147
    b. 2646
    c. 635,040
    d. 43,286

25. The number of different ways 8 children can be seated on a bench is

    a. 8
    b. 256
    c. 6720
    d. 40,320

26. A card is selected from an ordinary deck of 52 cards. The probability that it is a three given that it is a red card is

    a. $\frac{1}{13}$
    b. $\frac{2}{13}$
    c. $\frac{1}{4}$
    d. $\frac{1}{2}$

27. Three cards are selected from an ordinary deck of 52 cards without replacement. The probability of getting all diamonds is

    a. $\frac{1}{64}$
    b. $\frac{1}{12}$
    c. $\frac{3}{52}$
    d. $\frac{11}{850}$

28. A coin is tossed and a card is drawn from an ordinary deck of 52 cards. The probability of getting a head and a club is

    a. $\frac{1}{8}$
    b. $\frac{3}{4}$
    c. $\frac{1}{6}$
    d. $\frac{1}{4}$

29. When two dice are rolled, the probability of getting a sum of 5 or 7 is

    a. $\frac{1}{3}$
    b. $\frac{2}{3}$
    c. $\frac{5}{18}$
    d. $\frac{1}{4}$

30. The odds in favor of an event when $P(E) = \frac{3}{7}$ are

    a. 3 : 7
    b. 7 : 3
    c. 3 : 4
    d. 10 : 3

31. The odds against an event when $P(E) = \frac{5}{9}$ are

    a. 4 : 9
    b. 4 : 5
    c. 9 : 4
    d. 13 : 4

32. The probability of an event when the odds against the event are 4:9 are

    a. $\frac{9}{13}$

    b. $\frac{4}{9}$

    c. $\frac{5}{9}$

    d. $\frac{4}{13}$

33. A person selects a card at random from a box containing 5 cards. One card has a 5 written on it. Two cards have a 10 written on them, and two cards have a 3 written on them. The expected value of the draw is

    a. 4.5
    b. 1.8
    c. 3.6
    d. 6.2

34. When a game is fair, the odds of winning will be

    a. 1:2
    b. 1:1
    c. 2:1
    d. 3:2

35. A person has 2 pennies, 3 nickels, 4 dimes, and 1 quarter in her purse. If she selects one coin at random, the expected value of the coin is

    a. 4.7 cents
    b. 6.3 cents
    c. 8.2 cents
    d. 12.4 cents

36. The number of outcomes of a binomial experiment is

    a. 1
    b. 2
    c. 3
    d. Unknown

37. A survey found that one in five Americans say that he or she has visited a doctor in any given month. If 10 people are selected at random, the probability that exactly 3 visited a doctor last month is

    a. 0.101
    b. 0.201
    c. 0.060
    d. 0.304

38. A survey found that 30% of teenage consumers received their spending money from a part time job. If 5 teenagers are selected at random, the probability that 3 of them have income from a part time job is

    a. 0.132
    b. 0.471
    c. 0.568
    d. 0.623

39. A box contains 4 white balls, 3 red balls, and 3 blue balls. A ball is selected at random and its color is noted. It is replaced and another ball is selected. If 5 balls are selected, the probability that 2 are white, 2 are red, and 1 is blue is

    a. $\frac{72}{365}$
    b. $\frac{41}{236}$
    c. $\frac{52}{791}$
    d. $\frac{1}{14}$

40. If there are 200 typographical errors randomly distributed in a 500-page manuscript, the probability that a given page contains exactly 3 errors is

    a. 0.0063
    b. 0.0028
    c. 0.0072
    d. 0.0014

41. A recent study found that 4 out of 10 houses were undermined. If 5 houses are selected, the probability that exactly 2 are undermined is

    a. $\frac{4}{25}$
    b. $\frac{27}{125}$
    c. $\frac{216}{625}$
    d. $\frac{8}{25}$

42. The mean of the standard normal distribution is

    a. 1
    b. 100
    c. 0
    d. Variable

43. The percent of the area under the normal distribution curve that falls within 2 standard deviations on either side of the mean is approximately

    a. 68
    b. 95
    c. 99.7
    d. Variable

44. The total area under the standard normal distribution curve is

    a. 50%
    b. 65%
    c. 95%
    d. 100%

45. In the graph of the standard normal distribution, the values of the horizontal axis are called

    a. x values
    b. y values
    c. z values
    d. None of the above

46. The scores on a national achievement test are approximately normally distributed with a mean of 120 and a standard deviation of 10. The probability that a randomly selected student scores between 100 and 130 is

    a. 68.2%
    b. 34.1%
    c. 48.8%
    d. 81.8%

47. The heights of a group of adult males are approximately distributed with a mean of 70 inches and a standard deviation of 2 inches. The probability that a randomly selected male from the group is between 68 and 72 inches is

    a. 34.1%
    b. 68.2%
    c. 81.8%
    d. 95.4%

48. The average time it takes an express bus to reach its destination is 32 minutes with a standard deviation of 3 minutes. Assume the variable is normally distributed. The probability that it will take the bus between 30.5 minutes and 33.5 minutes to arrive at its destination is

    a. 38.3%
    b. 93.3%
    c. 34.1%
    d. 6.7%

49. The average time it takes pain medicine to relieve pain is 18 minutes. The standard deviation is 4 minutes. The variable is approximately normally distributed. If a randomly selected person takes the medication, the probability that the person experiences pain relief within 25.2 minutes is

    a. 50%
    b. 68.3%
    c. 75.2%
    d. 96.4%

50. The average electric bill for the month of October in a residential area is \$72. The standard deviation is \$5. If a resident of the area is randomly selected, the probability that his or her electric bill for the month of October is greater than \$75 is

    a. 27.4%
    b. 42.3%
    c. 72.6%
    d. 88.1%

Use the payoff table for Questions 51–55.

	Player B	
Player A	X	Y
X	−2	6
Y	5	−8

51. If player A uses Y and Player B uses X, the payoff is
    a. −2
    b. 5
    c. 6
    d. −8

52. The optimal strategy for Player A is to use X with a probability of

    a $\frac{8}{21}$

    b. $\frac{2}{3}$

    c. $\frac{13}{21}$

    d. $\frac{1}{3}$

53. The value of the game is
    a. $\frac{1}{3}$
    b. $\frac{13}{21}$
    c. $\frac{8}{21}$
    d. $\frac{2}{3}$

54. The optimal strategy for Player B would be to play Y with a probability of
    a. $\frac{8}{21}$
    b. $\frac{1}{3}$
    c. $\frac{13}{21}$
    d. $\frac{2}{3}$

55. The optimal strategy for Player B would be to play X with a probability of
    a. $\frac{1}{3}$
    b. $\frac{13}{21}$
    c. $\frac{2}{3}$
    d. $\frac{8}{21}$

Use the Period Life tables to answer questions 56–60.

56. What is the probability that a male will die at age 77?

    a. 0.058
    b. 0.031
    c. 0.025
    d. 0.018

57. What is the probability that a female age 50 will live to age 70?

    a. 0.016
    b. 0.473
    c. 0.837
    d. 0.562

58. What is the probability that a male age 65 will die before age 72?

    a. 0.747
    b. 0.339
    c. 0.661
    d. 0.176

59. What is the life expectancy of a 16-year-old female?

    a. 68.71 years
    b. 64.06 years
    c. 55.23 years
    d. 55.56 years

60. What is the median future lifetime of an 18-year-old male?

    a. 78 years
    b. 60 years
    c. 52 years
    d. 47 years

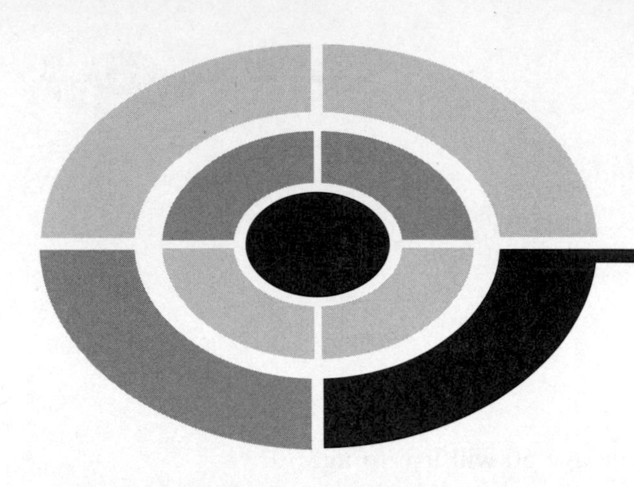

Answers to Quizzes and Final Exam

Chapter 1

1. d	6. d	11. c
2. b	7. a	12. d
3. b	8. b	13. b
4. a	9. c	14. c
5. a	10. b	15. d

Chapter 2

1. b	6. d	11. d
2. a	7. c	12. a
3. c	8. d	13. d
4. c	9. a	14. c
5. b	10. c	15. c

Chapter 3

1. b	6. b	11. a
2. c	7. a	12. d
3. a	8. c	
4. c	9. b	
5. d	10. b	

Chapter 4

1. c	6. c	11. c
2. a	7. a	12. a
3. b	8. c	13. b
4. d	9. b	14. d
5. b	10. b	

Chapter 5

1. a	6. a	11. b
2. c	7. a	12. c
3. b	8. d	13. d
4. b	9. c	14. b
5. b	10. b	15. d

Chapter 6

1. d	6. b	11. d
2. b	7. d	12. b
3. c	8. a	13. c
4. a	9. c	14. d
5. b	10. b	15. a

Chapter 7

1.	c	6.	c
2.	b	7.	a
3.	c	8.	b
4.	a	9.	c
5.	d	10.	a

Chapter 8

1.	c	6.	c	11.	b
2.	b	7.	a	12.	d
3.	d	8.	d	13.	c
4.	a	9.	b	14.	a
5.	d	10.	c	15.	c

Chapter 9

1.	b	6.	d	11.	b
2.	a	7.	a	12.	d
3.	b	8.	c	13.	d
4.	a	9.	b	14.	c
5.	c	10.	d	15.	b

Chapter 10

1. c
2. d
3. a
4. b
5. b

Chapter 11

1.	c	6.	c
2.	b	7.	c
3.	d	8.	d
4.	a	9.	b
5.	a	10.	c

Chapter 12

1.	d	6.	a
2.	b	7.	d
3.	a	8.	a
4.	c	9.	d
5.	c	10.	b

Final Exam

1.	d	13.	a	25.	d	37.	b	49.	d
2.	b	14.	d	26.	a	38.	a	50.	a
3.	a	15.	a	27.	d	39.	d	51.	b
4.	b	16.	c	28.	a	40.	c	52.	c
5.	a	17.	b	29.	c	41.	c	53.	d
6.	d	18.	a	30.	c	42.	c	54.	b
7.	b	19.	c	31.	b	43.	b	55.	c
8.	c	20.	b	32.	a	44.	d	56.	a
9.	a	21.	b	33.	d	45.	c	57.	c
10.	d	22.	d	34.	b	46.	d	58.	d
11.	c	23.	b	35.	c	47.	b	59.	b
12.	c	24.	b	36.	b	48.	a	60.	b

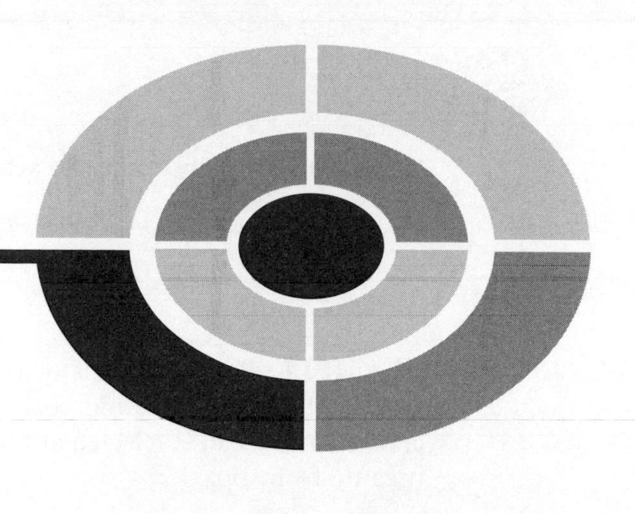

Appendix

Bayes' Theorem

A somewhat more difficult topic in probability is called Bayes' theorem.

Given two dependent events, A and B, the earlier formulas allowed you to find $P(A \text{ and } B)$ or $P(B|A)$. Related to these formulas is a principle developed by an English Presbyterian minister, Thomas Bayes (1702–1761). It is called *Bayes' theorem*.

Knowing the outcome of a particular situation, Bayes' theorem enables you to find the probability that the outcome occurred as a result of a particular previous event. For example, suppose you have two boxes containing red balls and blue balls. Now if it is known that you selected a blue ball, you can find the probability that it came from box 1 or box 2. A simplified version of Bayes' theorem is given next.

For two mutually exclusive events, A and B, where event B follows event A,

$$P(A|B) = \frac{P(A) \cdot P(B|A)}{P(A) \cdot P(B|A) + P(\overline{A}) \cdot P(B|\overline{A})}$$

EXAMPLE: Box 1 contains two red balls and one blue ball. Box 2 contains one red ball and three blue balls. A coin is tossed; if it is heads, Box 1 is chosen, and a ball is selected at random. If the ball is red, find the probability it came from Box 1.

SOLUTION:

Let A = selecting Box 1 and $\overline{A}$ = selecting Box 2. Since the selection of a box is based on a coin toss, the probability of selecting Box 1 is $\frac{1}{2}$ and the probability of selecting Box 2 is $\frac{1}{2}$; hence, $P(A) = \frac{1}{2}$ and $P(\overline{A}) = \frac{1}{2}$. Let B = selecting a red ball and $\overline{B}$ = selecting a blue ball. From Box 1, the probability of selecting a red ball is $\frac{2}{3}$, and the probability of selecting a blue ball is $\frac{1}{3}$ since there are two red balls and one blue ball. Hence $P(B|A) = \frac{2}{3}$ and $P(\overline{B}|A) = \frac{1}{3}$. Since there is one red ball in Box 2, $P(B|\overline{A})$ is $\frac{1}{4}$, and since there are 3 blue balls in Box 2, $P(\overline{B}|\overline{A}) = \frac{3}{4}$. The probabilities are shown in Figure A-1.

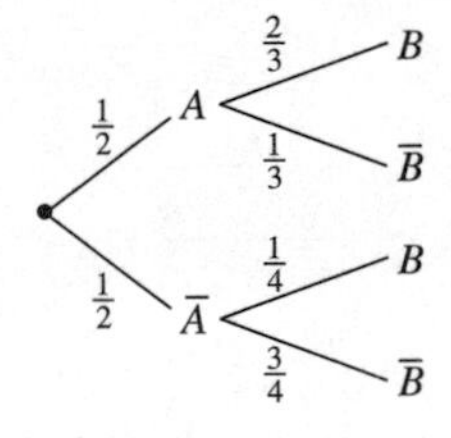

Fig. A-1.

Hence

$$P(A|B) = \frac{P(A) \cdot P(B|A)}{P(A) \cdot P(B|A) + P(\overline{A}) \cdot P(B|\overline{A})} = \frac{\frac{1}{2} \cdot \frac{2}{3}}{\frac{1}{2} \cdot \frac{2}{3} + \frac{1}{2} \cdot \frac{1}{4}} = \frac{\frac{1}{3}}{\frac{1}{3} + \frac{1}{8}} = \frac{8}{11}$$

In summary, if a red ball is selected, the probability that it came from Box 1 is $\frac{8}{11}$.

EXAMPLE: Two video products distributors supply video tape boxes to a video production company. Company A sold 100 boxes of which 5 were defective. Company B sold 300 boxes of which 21 were defective. If a box was defective, find the probability that it came from Company B.

SOLUTION:

Let $P(A)$ = probability that a box selected at random is from company A. Then, $P(A) = \frac{100}{400} = \frac{1}{4} = 0.25$; $P(B) = P(\overline{A}) = \frac{300}{400} = \frac{3}{4} = 0.75$. Since there are 5 defective boxes from Company A, $P(D|A) = \frac{5}{100} = 0.05$ and there are 21 defective boxes from Company B or $\overline{A}$, so $P(D|\overline{A}) = \frac{21}{300} = 0.07$. The probabilities are shown in Figure A-2.

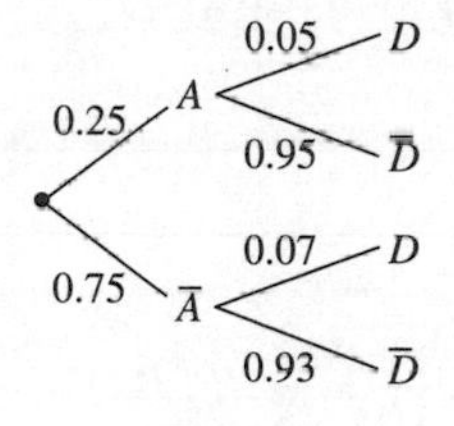

Fig. A-2.

$$P(A|D) = \frac{P(A) \cdot P(D|A)}{P(A) \cdot P(D|A) + P(\overline{A}) \cdot P(D|\overline{A})}$$

$$= \frac{(0.25)(0.05)}{(0.25)(0.05) + (0.75)(0.07)}$$

$$= \frac{0.0125}{0.0125 + 0.0525} = \frac{0.0125}{0.065} = 0.192$$

PRACTICE

1. Box I contains 6 green marbles and 4 yellow marbles. Box II contains 5 yellow marbles and 5 green marbles. A box is selected at random and a marble is selected from the box. If the marble is green, find the probability it came from Box I.
2. An auto parts store purchases rebuilt alternators from two suppliers. From Supplier A, 150 alternators are purchased and 2% are defective. From Supplier B, 250 alternators are purchased and 3% are defective.

Given that an alternator is defective, find the probability that it came from Supplier B.

3. Two manufacturers supply paper cups to a catering service. Manufacturer A supplied 100 packages and 5 were damaged. Manufacturer B supplied 50 packages and 3 were damaged. If a package is damaged, find the probability that it came from Manufacturer A.
4. Box 1 contains 10 balls; 7 are marked "win" and 3 are marked "lose." Box 2 contains 10 balls; 3 are marked "win" and 7 are marked "lose." You roll a die. If you get a 1 or 2, you select Box 1 and draw a ball. If you roll 3, 4, 5, or 6, you select Box 2 and draw a ball. Find the probability that Box 2 was selected if you have selected a "win."
5. Using the information in Exercise 4, find the probability that Box 1 was selected if a "lose" was drawn.

ANSWERS

1. $P(B1) = \frac{1}{2}$; $p(G|B1) = \frac{6}{10} = \frac{3}{5}$; $P(B2) = \frac{1}{2}$; $P(G|B2) = \frac{5}{10} = \frac{1}{2}$

$$P(B1|G) = \frac{P(B1) \cdot P(G|B1)}{P(B1) \cdot P(G|B1) + P(B2) \cdot P(G|B2)}$$

$$= \frac{\frac{1}{2} \cdot \frac{3}{5}}{\frac{1}{2} \cdot \frac{3}{5} + \frac{1}{2} \cdot \frac{1}{2}} = \frac{\frac{3}{10}}{\frac{3}{10} + \frac{1}{4}} = \frac{6}{11}$$

2. $P(A) = \frac{150}{400} = 0.375$; $P(D|A) = 0.02$; $P(B) = \frac{250}{400} = 0.625$

$P(D|B) = 0.03$

$$P(B|D) = \frac{P(B) \cdot P(D|B)}{P(B) \cdot P(D|B) + P(A) \cdot P(D|A)}$$
$$= \frac{(0.625)(0.03)}{(0.625)(0.03) + (0.375)(0.02)} = 0.714$$

3. $P(A)=\frac{100}{150}=\frac{2}{3}$; $P(D|A)=\frac{5}{100}=\frac{1}{20}$; $P(B)=\frac{50}{150}=\frac{1}{3}$

$P(D|B)=\frac{3}{50}$

$$P(A|D)=\frac{P(A)\cdot P(D|A)}{P(A)\cdot P(D|A)+P(B)\cdot P(D|B)}$$

$$=\frac{\frac{2}{3}\cdot\frac{1}{20}}{\frac{2}{3}\cdot\frac{1}{20}+\frac{1}{3}\cdot\frac{3}{50}}=\frac{\frac{1}{30}}{\frac{1}{30}+\frac{1}{50}}=\frac{5}{8}$$

4. $P(B1)=\frac{1}{3}$; $P(W|B1)=\frac{7}{10}$; $P(B2)=\frac{2}{3}$; $P(W|B2)=\frac{3}{10}$

$$P(B2|W)=\frac{P(B2)\cdot P(W|B2)}{P(B2)\cdot P(W|B2)+P(B1)\cdot P(W|B1)}$$

$$=\frac{\frac{2}{3}\cdot\frac{3}{10}}{\frac{2}{3}\cdot\frac{3}{10}+\frac{1}{3}\cdot\frac{7}{10}}=\frac{\frac{1}{5}}{\frac{1}{5}+\frac{7}{30}}=\frac{6}{13}$$

5. $P(B1)=\frac{1}{3}$; $P(L|B1)=\frac{3}{10}$; $P(B2)=\frac{2}{3}$; $P(L|B2)=\frac{7}{10}$

$$P(B1|L)=\frac{P(B1)\cdot P(L|B1)}{P(B1)\cdot P(L|B1)+P(B2)\cdot P(L|B2)}$$

$$=\frac{\frac{1}{3}\cdot\frac{3}{10}}{\frac{1}{3}\cdot\frac{3}{10}+\frac{2}{3}\cdot\frac{7}{10}}=\frac{\frac{1}{10}}{\frac{1}{10}+\frac{14}{30}}=\frac{3}{17}$$

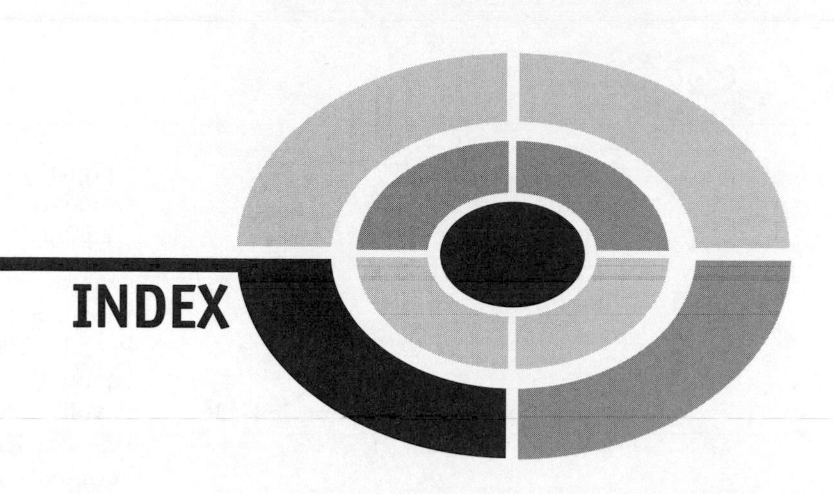

INDEX

Allan Bluman is Professor Emeritus of Mathematics at the South Campus of the Community College of Allegheny County, in Pennsylvania. He has taught most of the math and statistics courses on the campus, as well as arithmetic fundamentals, since 1972. Professor Bluman has written several articles and books, including *Modern Math Fun Book* (Cuisinaire Publishing) and *Elementary Statistics: A Step-by-Step Approach,* now in its Fifth Edition, and *Elementary Statistics: A Brief Version,* now in its Second Edition, both from McGraw-Hill.